Ahlem LAISSOUF

Effect of flax on obesity and biological aging

Ahlem LAISSOUF

Effect of flax on obesity and biological aging

Linseed oil and obesity

ScienciaScripts

Cover image: www.ingimage.com

This book is a translation from the original published under ISBN 978-3-8381-7708-3.

Publisher:
Sciencia Scripts
is a trademark of
Dodo Books Indian Ocean Ltd. and OmniScriptum S.R.L publishing group

120 High Road, East Finchley, London, N2 9ED, United Kingdom
Str. Armeneasca 28/1, office 1, Chisinau MD-2012, Republic of Moldova, Europe
Managing Directors: Ieva Konstantinova, Victoria Ursu
info@omniscriptum.com

Printed at: see last page
ISBN: 978-620-8-41258-6

Contents

ACKNOWLEDGEMENTS

First of all, I would like to express my most sincere gratitude to my thesis supervisor, Professor **SOULIMANE-MOKHTARI Nassima.** I am also grateful to her for the considerable time she me, her teaching and scientific qualities, her frankness and her sympathy. I learned a lot from her and I am very grateful to her for that. Her support was crucial in bringing this work to a successful conclusion.

If I have managed to complete my thesis today, it is thanks to his guidance and encouragement. I would like to express my admiration for his great qualities, both scientific and human. Many thanks for your support, your advice, your teaching and your unshakeable confidence.

I would like to express my gratitude and respect to the members of the jury.

I would like to begin by thanking Mrs **MERZOUK Hafida**, University Professor and Director of the Physiology, Physiopathology and Biochemistry of Nutrition Research Laboratory, who agreed to devote her time to examining and judging this work as Chairman of the Jury. She spared no effort to ensure that this work was carried out in the best possible conditions. This work is the fruit of your inspiration. I would like to express my deep gratitude and thank you for giving me opportunity and privilege to work alongside you in your laboratory.

I would like to thank Professor **BOUANANE Samira** for the honour she has done me by agreeing to judge this thesis and to be an examiner.

I would also like to express my gratitude to Professor **AIT YAHIA Dalila**, who agreed to sit on the jury as an examiner.

I would also like to thank Professor **SLIMANI Miloud**, who did me the honour examining my work and taking part in the jury.

I would also like to express my sincere thanks to Professor **AOUES Abdelkader** for the honour he has done me by agreeing to judge this work.

I would also like to express my sincere gratitude to **Mr MERZOUK Sid Ahmed**, for his dedication and interest in my research project and his scientific support throughout this work, his involvement, his constructive criticism in the field of statistics and his encouragement, all of which contributed to the production of a work.

I would also like to thank the team at the UPRES lipids laboratory, Gabriel Faculty of Science, University of Burgundy, Dijon, France.

Finally, I would like to thank all those who contributed one way or another to the completion of this thesis.

SUMMARY

Obesity is associated with metabolic disorders and intense oxidative stress, which can be aggravated with age. The aim of this study was to test the beneficial effect of linseed oil, which is rich in n-3 polyunsaturated fatty acids (n-3 PUFAs), on disorders of lipid metabolism and redox status induced by the cafeteria diet (hyperlipidic and hypercaloric) in aged wistar rats. The aged rats were divided into six groups fed the control (standard) diet with or without linseed oil at (2.5% or 5%) and others fed the cafeteria diet with or without linseed oil at both concentrations. The rats were sacrificed after two months on the diet. Blood and organ samples (liver, adipose tissue, muscle, intestine) were taken to determine changes in lipid metabolism and redox status. The results show that the cafeteria diet induces obesity, hyperglycaemia and hyperlipidaemia with altered serum and organ fatty acid composition in aged wistar rats. These rats also showed altered redox status, with increased plasma and tissue levels of malondialdehyde (MDA), carbonylated proteins (PCAR) and markers of lipoprotein oxidation, reduced vitamin C and altered catalase and glutathione activities, with a reduction in plasma and an increase in liver and muscle, indicating clear oxidative stress. Supplementation of the diet with flaxseed oil *results in* a reduction in body weight, a reduction in glycaemia and in serum cholesterol and triglyceride levels and lipoproteins, and modulates the enzymatic activities involved in lipid metabolism (LPL, LCAT, LHS). The fatty acid composition of liver, adipose tissue and serum is also corrected. Oxidant/antioxidant status showed an improvement, characterised by an increase in the activities of antioxidant enzymes and vitamin C and a decrease in the levels of MDA, hydroperoxides and carbonylated proteins and markers of lipoprotein oxidation. In conclusion, alterations in lipid metabolism and redox status are exacerbated by obesity during ageing. Flaxseed oil improves these disorders thanks to its high content of (PUFA-In-3). Therefore, early prevention through appropriate nutrition is necessary to limit the development of chronic diseases.
Key words: obesity, aged rats, cafeteria diet, oxidant/antioxidant status, PUFAs.

ملخص

ترتبط السمنة بعدة تعقيدات أيضية تزداد سوءا مع التقدم في السن. تهدف هذه الدراسة أساسا إلى اختبار التأثيرات الايجابية لزيت الكتان الغني بالأحماض الذهنية المتعددة غير المشبعة على أيض الدهون ,البروتينات ,السكريات وأيضا الوضع التأكسدي الذي يسببه النظام الغذائي كافتيريا (عالية الدهون و السعرات الحرارية) على الفئران المسنة من نوع ويستار. تم تقسيم الفئران المسنة إلى ستة مجموعات اتبعت نظاما غذائيا عاديا أو حمية كافتيريا المضاف له أو لا زيت الكتان بنسبة 2,5%(أو5%) بعد شهرين من النظام الغذائي تم إجراء التضحيات على الفئران وفحص عينات الدم و الأعضاء المتمثلة في (الكبد, الأنسجة الذهنية ,العضلات و الأمعاء) وذلك من أجل تحديد التغيرات الحاصلة في أيض ا لدهون وحالة الأكسدة. أظهرت النتائج أن النظام الغذائي كافتيريا يؤدي إلى الإفراط في الأكل مما يسبب الزيادة في وزن الجسم والنسيج الشحمي وتشبعه بالدهون وأيضا إلى اضطرابات في الوضع التأكسدي. استهلاك غذاء غني بزيت الكتان يؤدي إلى تخفيض وزن الجسم, انخفاض نسبة السكر في الدم وكذلك الكولسترول والدهون الثلاثية والبروتينات الذهنية و تنظيم أنشطة إنزيمات التمثيل الغذائي للدهون (LPL,LCAT,LHS) إضافة إلى تصحيح تكوين الأحماض الذهنية في المصل ,الكبد، والأنسجة الذهنية و تحسن الوضع التاكسدي من خلال زيادة أنشطة الإنزيمات المضادة للأكسدة، وفيتامين C و انخفاض النسب البلازمية والنسيجية لل PCAR, MDAو علامات أكسدة البروتين الذهني ختاما إظطرابات أيض الدهون وحالة الأكسدة تزداد مع السمنة و التقدم في السن. زيت الكتان الغني بالأحماض الذهنية المتعددة غير المشبعة يخفف من هذه الاضطرابات و عليه فإن الوقاية المبكرة من خلال التغذية المناسبة على وجه الخصوص يعد أمرا ضروريا للحد من تطور الأمراض المزمنة.

الكلمات المفتاحية: السمنة, الفئران المسنة , حمية كافتريا, الوضع التاكسدي, الأحماض الدهنية المتعددة غير المشبعة

DEDICACES

With the help of God, the all-powerful, merciful and forgiving, I have been able to complete this work, which I dedicate to :

To my dear mother

As a token of my deep gratitude and undeniable appreciation, for all the sacrifices she makes for me, all the trust she places in me and all the love with which she surrounds me.

To my dear father

In expressing my gratitude, my deep love and my passion, for his trust, his moral and material support and for his infinite love.

To my husband Amar

To my son Mustapha Anés and my in-laws

To Hanane, Assia , Amina , Dalel and Mamia

*To all my **family** and friends*

*To all **the teachers** who have contributed to my training.*

To all those who contributed one way or another to the development of this work.

SCIENTIFIC PUBLICATIONS

LAISSOUF A, MOKHTARI-SOULIMANE N, MERZOUK H, BENHABIB N (2013). Dietary flaxseed oil supplementation improves the oxidant antioxidant status in obese aged rats. ***International journal of medicine and pharmaceutical sciences***, 3(2): 87-94.

SCIENTIFIC COMMUNICATIONS

1 er Congres International De La Société Algérienne de Nutrition. 05-06 December, 2012, Oran, Algeria.

LAISSOUF A, KHOLKHAL F, AYAD A, MOKHTARI SOULIMANE N, MERZOUK H (2012). The effects of flaxseed oil enriched cafeteria diet on oxidant/antioxidant status in aged wistar rats.

International Congress IWIB4 10-11April 2013 Tlemcen, Algeria.

LAISSOUF A, MOKHTARI N, MERZOUK H (2013).Therapeutic effect of linseed oil (*Linum usitatissimum)* on lipoprotein lipase and hyperglycemia in the management of obesity.

3rd International Congress on Bioactive Molecules, Functional Foods and Diseases Associated with Oxidative Stress .21-23 March, 2014, Hammamet, Tunisia

LAISSOUF A, MOKHTARI N, MERZOUK H (2014). The effect of cafeteria diet enriched with linseed oil "*linum usitatissimum*" on oxidative stress in aged wistar rats.

National Congress IVth Scientific Days of the Faculty of Natural and Life Sciences, 9-10 April 2013 Mostaganem, Algeria

LAISSOUF A, MOKHTARI N, MERZOUK H (2013). The effect of "Linum usitatissimum" flaxseed oil enriched cafeteria diet on body weight and lipid metabolism in wistar aged rats.

First seminar in engineering, health and analysis (SEHA) 05 may 2013, Alger, Algérie.

LAISSOUF A, MOKHTARI SOULIMANE N, MERZOUK H (2013). The therapeutic effect of flaxseed oil (Linum Usitatissimum) on hyperglycemia and hypercholesterolemia in obese wistar rats.

Forum on the development of life sciences and the universe 14-15 May 2013, Tlemcen, Algeria.

LAISSOUF A, MOKHTARI N, MERZOUK H (2013). The effect of "Linum usitatissimum" flaxseed oil enrichment on plasma and tissue malodialdehyde in obese wistar rats.

INTRODUCTION

Nutrition is one of the major factors contributing to the onset of various diseases.

It is not the only cause of these pathologies, but it is an essential contributing factor among other environmental or genetic factors. It is a factor in which it is possible to intervene.

The issue of excess weight has been identified as one of the major public health problems of the 21st century (WHO, 2000), due to its potential impact on health and its increasing frequency (STURM, 2007; JAMES, 2008; DE SAINT POL, 2009). Over the last few decades, the incidence of obesity has risen dramatically to the point of becoming a truly global epidemic (KOPELMAN, 2000; CABALLERO, 2007), affecting the majority of nations, regardless of their level of development (BRANCA, 2008).

Obesity reflects a failure of the energy reserve regulation system due to external (lifestyle, environment) and/or internal (psychological or biological, in particular genetic and neuro-hormonal) factors (BASDEVANT and GUY-GRAND, 2004). The obesity epidemic is largely due to the westernisation of the diet (FRANCIS et al., 2009; KELLI et al., 2009). Indeed, in its latest recommendations, the WHO points out that "poor diet is a risk factor for non-communicable diseases and contributes to overweight and obesity" (WHO, 2010). This westernisation is accompanied by a high fat intake, and fat is the most calorific food source, and excess fat is stored as triglycerides in adipocytes (MICHALIK et al., 2000; FRANCIS et al., 2009). The adoption of a high-fat and/or high-carbohydrate diet is the main cause of excess weight. In both humans and animals, studies show a relationship between excess calories (most often provided by excess lipids) and an increase in body fat (WEST and YORK, 1998; AILHAUD, 2007).

Adipose tissue plays an essential role in the storage and mobilisation of energy, thus contributing to the regulation of lipid and lipoprotein metabolism. Excess adipose tissue can alter lipid and lipoprotein metabolism by modifying their plasma levels and composition, with very significant health consequences (DIXON, 2010). Obesity is associated with various pathological conditions, including dyslipidaemia, impaired glucose tolerance and type II diabetes. Furthermore, in patients diagnosed with metabolic syndrome, arterial hypertension and coronary artery disease are frequent, considerably increasing the subsequent risk of myocardial infarction (GRUNDY, 2006). Obesity is associated with increased mortality, particularly cardiovascular mortality (WHITLOCK et al., 2009). Given that, on the one hand, the development of cardiovascular disease in obesity is the result of a constellation of pro-atherogenic mechanisms, and, on the other hand, that alterations in lipid and lipoprotein metabolism are a major cause of the development and progression of cardiovascular disease (ALBERTI et al., 2005). During ageing, the same anomalies are observed and even accentuated (KELLEY et al., 2005).

It has also been established that the accumulation of fat in adipose tissue during obesity leads to the induction of systemic oxidative stress in rodents and humans (GOURANTON and LANDRIER, 2007). Oxidative stress corresponds to a state of imbalance in the redox balance in biological systems. This imbalance may be due to an overproduction of reactive oxygen species (ROS), a reduction in antioxidant defences or a combination of these two phenomena (FULOP et al., 2010). Once formed, ROS can cause oxidative damage, often irreversible, to a large number of biological substrates (enzymes, proteins, DNA, lipids, glucose). This is the cause of numerous pathologies, including type II diabetes (MAIESE CHONG et al. 2007), obesity (VINCENT, INNES et al., 2007) and ageing (ROMANO and SERVIDDIO, 2010). The involvement of free radicals and oxidative stress in ageing was proposed over 50 years ago (HARMAN, 1956). With age, the frequency of degenerative pathologies linked to the ageing process increases. Research over the last few decades has shown that a large number of these pathologies, including cancer, cardiovascular disease, dementia, cataracts and a decline in immune function, are promoted by the production of free radicals (FULOP et al., 2010).

While nutrition is a risk factor, it can also be a protective factor. A number of studies have highlighted the benefits of n-3 polyunsaturated fatty acids, particularly with regard to dyslipidemia and oxidative stress (LINCHTENSTEIN et al., 1998; YESSOUFOU et al., 2006). Indeed, it would appear that the countries least affected by obesity are Korea and Japan, which are major consumers of fish (very rich in polyunsaturated fatty acids) (LEE et al., 2002). The Mediterranean diet (based on vegetables, fruit and cereals) also reduces the development of cancer because it is rich in PUFAs, provided by fish and olive oil (LAVECCHIA, 2004). Most studies on the effect of PUFAs on the development of adipose tissue in the case of obesity have focused on n-3 PUFAs (BUCKLEY et al., 2010).

Flaxseed is one of the richest oils in n-3 PUFAs. The benefits of linseed oil can linked to the potential of α-linolenic acid (ALA). According to SIMOPOULOS and ROBINSON (1998), linseed oil is one of the most highly unsaturated of all oils. It represents the richest vegetable source of alpha-linolenic acid (ALA; 50- 62% linseed oil), or (22% whole linseed) (PAN et al., 2009). In addition, linseed oil contains 12.7% linoleic acid, giving it the highest n-3/n-6 ratio of all vegetable sources (TZANG et al., 2009).

Studies have reported a reduction in visceral fat mass in animals fed an obesogenic diet enriched with ALA, compared with diets rich in saturated fatty acids or linoleic acid (LA) (CHICCO et al., 2009; BARANOWSKI et al., 2012).

Data obtained in humans and animals suggest that n-3 PUFAs are capable of reducing lipogenesis (DUPLUS et al., 2002 ; FLACHS et al., 2009) and of stimulating the oxidation (mitochondrial and peroxisomal) of fatty acids (BUCKLEY et al., 2010 ; ROBINSON et al., 2007).

Numerous studies have also shown that a diet enriched with n-3 PUFAs induces a variation in energy balance and body weight, leading to a reduction in obesity (LORENTE et al., 2013). Other studies in rats have shown that fish oil has a beneficial effect on insulin resistance (STORLIEN et al., 1987). Studies in humans and experimental models have also shown that n-3 PUFAs reduce plasma and liver triglyceride levels (HARRIS, 1989; LICHTENSTEIN et al., 1998), as well as cholesterol levels by increasing biliary excretion (ZHAO et al., 2004). In addition, the beneficial effects of n-3 PUFAs on oxidative stress appear to be increasingly evident. n-3 PUFAs modulate the activities of antioxidant enzymes and increase the effectiveness of the body's antioxidant defence system (DEMOZ et al., 1992; VENKATRAMAN et al., 1994).

The combination of obesity and ageing can accentuate the state of oxidative stress and may alter the state of health. This is the context of our work, which focuses on the benefits of flaxseed oil, which is rich in n-3 PUFAs, on lipid metabolism disorders and antioxidant oxidative status linked to obesity and age.

This work is based on the use of an experimental model of nutritional obesity. Over a two-month experimental period, the wistar rat is fed a high-fat, high-calorie diet, known as the cafeteria diet, which is rich in saturated fatty acids. The components of this diet were chosen to mimic the eating habits observed in humans. Initially, this diet was fed to aged rats in comparison with the standard control diet, to gain a better understanding of the impact of overfeeding (cafeteria effect) in aged wistar rats. Next, the cafeteria diet and the control diet were supplemented with 2.5 and 5% linseed oil (Linum usitatissimum oil) and given to other groups of aged rats to determine their benefits on metabolism and oxidant/antioxidant status.

CHAPTER 1

CURRENT STATUS OF THE SUBJECT

I. Obesity, ageing and dyslipidemia

In recent decades, our eating habits have undergone profound changes in line with the transformations in society brought about by globalisation, industrialisation and urbanisation (AILHAUD, 2007). These changes correspond to a nutritional transition characterised by westernised dietary profiles (abundance of fatty and sugary foods, industrial food products rich in energy but with low nutritional density) and sedentary lifestyles. This nutritional transition is an important factor in the genesis of obesity, in association with other behavioural factors (meal destructuring, snacking, binge eating, consumption of fast food and fizzy drinks, etc.) and genetic factors (JUNIEN et al., 2005).

Since the late 1990s, obesity has reached the stage of a global epidemic, mainly affecting industrialised and developing countries. More than 1.5 billion adults worldwide were overweight in 2008, of whom 502 million were obese. By 2030, the number of overweight people is expected to reach 3.3 billion. Algeria, like other Maghreb countries, has not been spared this modern-day scourge (Kemali, 2003). A study of the body mass index (BMI) shows that 15% of the Algerian population is obese (KEMALI, 2003).

Obesity is one of the risk factors for the onset of numerous pathologies, such as cardiovascular disease and type 2 diabetes, increasing the risk of morbidity and mortality in both industrialised and developing countries (SAKURAI, 2000 ; WHITLOCK et al., 2009).

Obesity is defined as an increase in body fat, with consequences for physical, psychological and social well-being. Human obesity reflects a failure of the energy reserve regulation system due to external (lifestyle, environment) and/or internal (psychological or biological, particularly genetic and neuro-hormonal) factors (BASDEVANT and GUY-GRAND, 2004). In the majority of cases, fat inflation is due to an inability to cope with excess food intake and insufficient energy expenditure (Figure 1). This imbalance can be accentuated by an increase in storage capacity. There are therefore four pathophysiological players: food intake, energy expenditure, adipose tissue and communication between the organs involved in controlling the energy balance (BASDEVANT, 2006).

With regard to the relationship between dietary intake and weight gain, interest has focused on the respective roles of fat and carbohydrate intake (LISSNER and HEITMANN, 1995; LECERF, 2008). Dietary fats have a higher energy value than other macronutrients, low satiety, high energy density and high palatability, which may explain why a high-fat diet can lead to an increase in energy intake, a phenomenon known as "passive overeating" or "fat bingeing" (BLUNDELLE and KING, 1996). The body tries to compensate for the excess energy consumption caused by fat-rich foods, but the post-ingestive appetite control signals triggered by fat ingestion are too weak or delayed, as digestion and absorption of fat is slow: they cannot prevent the rapid consumption of a high-fat meal. A meal rich in fat leads to energy intake around 10% higher than a meal rich in carbohydrates, and despite this higher intake, the food intake of the following meal is not reduced. Control of lipid intake is regulated in the long term, while the balance of lipid substrates appears poorly regulated in the short term, and the energy expenditure associated with their use (intestinal absorption, processing, storage) is low. Postprandial thermogenesis and energy expenditure linked to storage represent only 4% of the energy ingested for lipids (30% for proteins, 9% for carbohydrates). Because of the low oxidative autoregulation capacity of fat substrates, excess fat intake is largely stored in adipose tissue, leading to an increase in fat mass over the long term.

In a large number of cross-sectional studies on different populations, a positive association has been found between fat intake (as a % of total energy intake) and indicators of obesity (usually BMI) (LISSNER and HEITMANN, 1995).

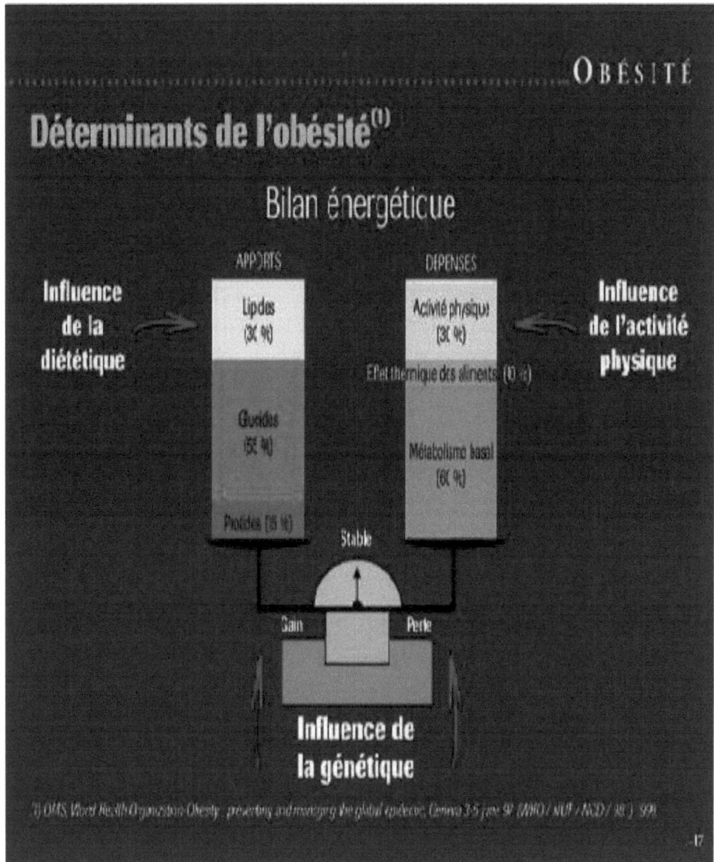

Figure 1: Determinants of obesity (WHO, 1999)

Adipose tissue is the body's main reservoir of mobilisable energy (KIESS et al., 2008). In mammals, there are two types of adipose tissue: white adipose tissue , which is the largest energy reserve, and brown adipose tissue, which is primarily involved in thermogenesis (SAELY et al., 2012). The advantage of this form of energy reserve is that it is particularly cost-effective in terms of density, due to the hydrophobic nature of lipids. It is also a thermal and mechanical protection system. Finally, the adipocyte is an endocrine and paracrine organ.

Adipocytes secrete a large number of substances (Figure 2), which influence energy balance, immune function, hormonal status and metabolism. Adipose tissue receives and sends a series of signals to its metabolic partners (HAUSMAN et al., 2001). The main substances include leptin, adiponectin, TNF- α, plasminogen activator inhibitor I (PAI-1), angiotensinogen, estradiol, etc., but many other molecules are also produced (AILHAUD, 2002).

Adipose tissue has exceptional plasticity (PENICAUD et al., 2000) and remains capable of developing. The increase in fat mass results from an increase in the size of the adipocytes (hypertrophy) or their number (hyperplasia), or both. Hypertrophy generally precedes hyperplasia. Hypertrophy results from an accumulation of triglycerides. Adipocyte size is the result of the lipogenesis/lipolysis balance. Beyond a certain size, the adipose cell no longer grows; the increase in storage capacity requires an increase in the number of cells. This is known as hyperplasia, and the number of adipocytes can increase dramatically. The increase in the number adipocytes results from the process of adipogenesis, which involves the proliferation of stem cells and their differentiation into adipocytes. Many intrinsic or extrinsic, molecular and cellular factors are involved the proliferation of adipose tissue (GAUVREAU et al., 2011).

This complex process is controlled by various signals that modify activity of transcriptional factors. The two main families involved are CCAAT/enhancer binding proteins (C/EBP) and peroxisome proliferator-activated receptors (PPAR), which belong to the super-family of steroid-type nuclear hormone receptors. According to the 'critical size' hypothesis, there is a maximum cell size. In this way, the differentiated adipose cell becomes loaded with triglycerides until it reaches a critical size, beyond which it 'recruits' a new pre-adipocyte. This can lead to an increase in the number of adipocytes, i.e. hyperplasia (GAUVREAU et al., 2011).

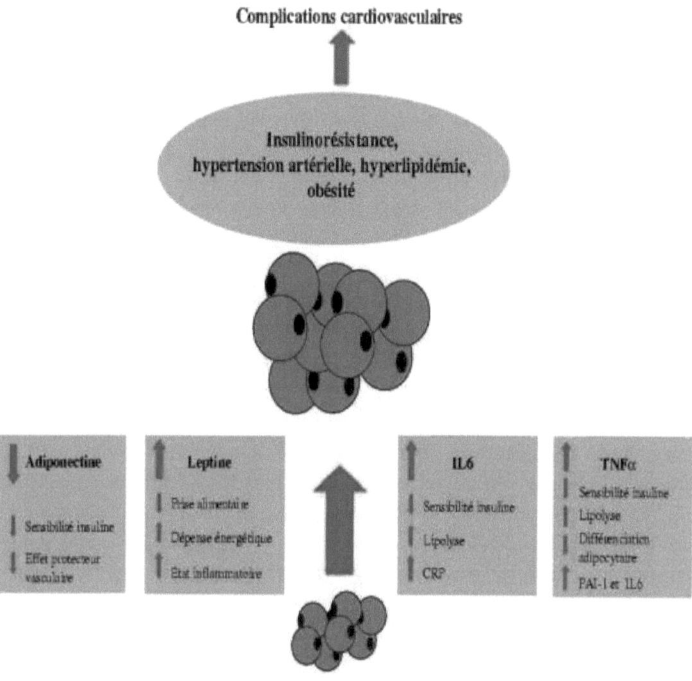

Figure 2. The development of adipose tissue is accompanied an increase in the secretion of pro-inflammatory adipokines which may contribute to the complications of obesity (FRUHBECK et al., 2001).

IL6: interleukin 6
PAI-1: plasminogen activator inhibitor 1
CRP: C-reactive protein
TNFa: tumour necrosis

Obesity leads to alterations in carbohydrate and lipid metabolism, leading to chronic complications including dyslipidaemia, which is the major cause of atherosclerosis and hence cardiovascular disease (CHAPMAN and SPOSITO, 2008). Dyslipidaemia or hyperlipoproteinaemia are common conditions involving a permanent increase in the plasma concentration one or more classes of lipoproteins. Atherosclerosis, a consequence of dyslipidaemia, is attributable to abnormalities in the metabolism of lipids and plasma lipoproteins. Obesity is accompanied by high plasma levels of triglycerides, free fatty acids and cholesterol, as well as an increase in VLDL and LDL and a reduction in serum HDL (REAVEN, 2005; GRUNDY, 2006). Hyperinsulinism, observed in obesity, is characterised by increased lipogenesis and/or reduced lipolysis (BJOMTORP, 1991). An excess of fat mass is also noted in obese rats, whose adipocytes, whose number remains unchanged, are rich in lipids and are larger in size than the adipocytes of control rats. These become less sensitive to the action of insulin, which reduces their capacity to convert glucose into free fatty acids and triglycerides. This reduction is considered to be a lipogenesis feedback control mechanism in adipose cells that are already very rich in lipids (GELARDI et al., 1990). In addition, qualitative and quantitative changes in lipids and lipoproteins involving changes specific enzymatic activities (LHS) have been noted in the obese (JOCKEN, 2007).

During ageing, the same anomalies are observed and even accentuated (KELLEY et al., 2005). Furthermore, ageing is a process that continues to fascinate biologists from all horizons, whether they interested in evolution, genetics, signalling or environmental toxicity. Ageing is characterised by progressive pathophysiological deterioration leading to homeostatic alteration, with a progressive decline in physiological capacity, a reduction in the ability to respond adaptively to environmental stimuli with age, and increased susceptibility to disease (DEL VALLE, 2011).

Ageing is associated with disturbances in carbohydrate metabolism in a significant population of elderly subjects. Tolerance to a glucose load is reduced in elderly people without diabetes mellitus or obesity, indicating a degree of insulin resistance. These changes contribute to an increase in blood glucose levels in postprandial or stress situations, which may increase the processes of non-enzymatic glycation of proteins (SZOKE et al., 2007).

Lipid metabolism is also disturbed. The increase in cholesterol levels around age of 50 is due to an increase in LDL due to a fall in its clearance. Hypertriglyceridaemia corresponds to an increase in VLDL, whose clearance is reduced by a drop in lipoprotein lipase activity.

In long-term nutritional studies, animal models remain irreplaceable. Several animal models have been used to investigate chronic diseases that threaten public health, such as obesity, type 2 diabetes, hypertension, etc. Among these models, the ob/ob mouse and the fa/fa Zuker rat are the most widely used (CARROLL et al., 2004; RUTH et al., 2008). Several types of experimental animal diet have been described to study the effect of nutritional factors. Most of these diets focus on one nutrient or category of nutrient (e.g. lipids, fructose, sodium) but this does not represent the westernised human diet (DEMIGNE et al., 2006). The first description of a high-fat diet (HFD) inducing nutritional obesity was described in 1959 by MASEK and FABRY. Subsequent studies have shown that this diet induces hyperglycaemia and insulin resistance (BUETTNER et al., 2006). In addition, this type of diet is characterised not only by its ability to induce increased lipid storage in adipose tissue, but also by an increase in oxidative stress throughout the system (SCOARIS et al., 2010). In addition to the hyperfat diet, nutritional obesity is induced by the cafeteria diet. This diet includes a variety of calorie-rich, palatable foods consumed by humans, such as crisps, chocolate, pâté, sausages, cheese, biscuits, etc., in varying proportions. This high-calorie, high-fat diet induces hyperphagia and obesity in rats (LOUIS-SYLVESTER, 1984). This model is similar to the development of nutritional obesity in humans following the voluntary over-consumption of these tasty foods (BARBER et al., 1985).

The highly palatable, calorie-rich cafeteria diet can alter the animal's energy balance. The increase in food intake observed in studies of obesity in humans can be explained by a strong appetite for sweet, high-fat foods. However, rats fed the cafeteria diet, which has free access to its various components, also choose sweet, high-fat foods (ESTEVE et al., 1994). Typically, this diet induces hyperphagia, increased food intake and obesity (SHAFAT et al., 2009). Furthermore, rats fed the cafeteria diet show metabolic abnormalities and obvious oxidative stress (BOUANANE et al., 2009). Thus, the use of obesity models based on the consumption of a diet enriched in fat and/or sugar seems to be relevant for reproducing the obesity commonly observed in humans. In order to reproduce Western human eating habits as closely as possible, we therefore chose use a 'cafeteria' type diet consisting of a mixture of commercial foods rich in fat and sugar (GROUBET et al., 2003).

II. Obesity, ageing and oxidant/antioxidant status

Our bodies constantly produce reactive oxygen species (ROS), also known as free radicals, as a result of the oxidative metabolism of oxygen. Free radicals are chemical species (atoms or molecules) that have one or more single electrons (unpaired electrons) on their outer layer and are capable independent existence (HALLIWELL et al., 1989). They can be derived from oxygen or other atoms such as nitrogen. The presence of a single electron gives free radicals a high degree of reactivity (short half-life) and they can be both oxidising and reducing species. These free radicals include the superoxide anion, the hydroxyl radical, singlet oxygen and hydrogen .

ROS are present in the cell in reasonable doses, and their concentration is regulated by the balance between their rate of production and their rate elimination by antioxidant systems (HALLIWELL et al., 1989). Thus, in the quiescent state, the antioxidant/pro-oxidant balance (redox balance) is said to be in equilibrium. However, this redox homeostasis can be disrupted, either by excessive ROS production (as in ageing or atherosclerosis), or by a reduction in antioxidant capacity (as in people suffering from obesity). This is known as oxidative stress (SIES, 1991). The paradox of ROS is that they are potentially toxic products of metabolism and are at the same time essential molecules in cell signalling and regulation (Figure 3).

ROS play an important physiological role by acting in low concentrations as secondary messengers capable of :

- regulate the phenomenon of apoptosis, which is the programmed suicide of cells progressing towards cancer (CURTIN et al., 2002)

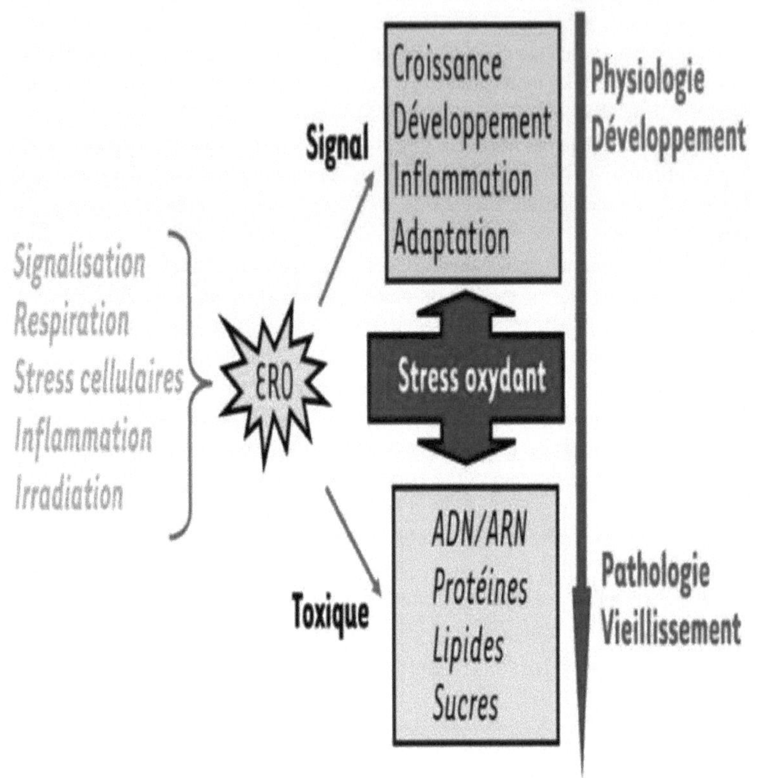

Figure 3: The double life of reactive oxygen species (ROS): contradictory pleiotropy and compromise (DELATTRE, 2005)

- activate transcription factors responsible for activating genes involved in the immune response (OWUOR and KONG, 2002)
- Modulate the expression of structural genes encoding antioxidant enzymes (HOLGREM, 2003)

If too many free radicals are produced, they can cause major cellular damage (CURTIN et al., 2002), by causing mutations in deoxyribonucleic acid (DNA) (CADET et al., 2002; HOURIGAN, 2010), by inactivating proteins, or by inducing lipid peroxidation processes in the polyunsaturated fatty acids of lipoproteins or the cell membrane. Lipid peroxidation results from the attack of free radicals on a double bond of a polyunsaturated fatty acid, whether unesterified or esterified (cholesterol esters, phospholipids, etc.). In the presence of molecular oxygen, chain reactions follow, giving rise to numerous compounds with different structures, such as aldehydes, isoprostanes, etc. (MOORE et al., 1998; DE ZWART et al., 1999). Aldehydes, including malonic dialdehyde (MDA) derived from the decomposition of hydroperoxides, can react with thiobarbituric acid to give coloured or fluorescent compounds. This reaction is based on an assay (TBARS: thiobarbituric acid-reactive substances). Several studies have been carried out in humans to determine whether there is a relationship between the plasma concentration of TBARS and ageing. In a study carried out on the plasma of 66 "normal" subjects aged between 19 and 90, researchers observed a highly significant correlation between TBARS concentration and the age of the subjects (POUBELLE et al., 1982).

Several studies have been carried out to determine whether ageing leads to an accumulation of carbonyl groups in proteins at the cellular level. This was measured in cultured human fibroblasts as a function of the donor's age. It was clearly demonstrated that this content increases exponentially with the age of the donor (STADTMAN, 1992).

The cellular origins of ROS are essentially enzymatic and derive from several sources. There are two main sources. The first results from imperfections in the mitochondrial respiratory chain, which produces ROS by monoelectronic reduction. The second major source of ROS production is NAD(P)H oxidase, essentially located in the plasma membrane, which interacts with the intracellular substrate (NADH,H+, or NADPH,H+) and releases the superoxide ion preferentially outside or inside the cell (BEAUDEUX et al., 2005). In addition to these two major sources of ROS, other cytosolic sources or sources present in various cellular organelles may play a role in modulating cell signalling, such as xanthine oxidase, enzymes of the smooth endoplasmic reticulum (cytochromes P450), NO synthases and enzymes of the arachidonic acid pathway. Thc main sourccs of rcactivc oxygen and nitrogcn spccics arc shown in Figure 4.

The balance between the positive and negative effects of free radicals is therefore particularly delicate. The production of free radicals is strictly regulated by our bodies, which have developed antioxidant defences to protect us against the potentially destructive effects of free radicals. These systems consist of (LEVINE and KIDD, 1996; GHISELLI et al., 2000)

- Enzymes (superoxide dismutases, catalase, glutathione peroxidases, thioredoxin-thioredoxin reductase pairs, heme oxygenase)
- Iron and copper transporter proteins (transferrin, ferritin, ceruleoplasmin)
- Small antioxidant molecules (glutathione, uric acid, bilirubin, glucose, vitamin A, C, E, ubiquinone, carotenoids, flavonoids).
- Trace elements (copper, zinc, selenium) essential for the activity of antioxidant enzymes.

A secondary defence system consisting of enzymes whose role is to prevent the accumulation of oxidised proteins or DNA in the cell and to degrade their toxic fragments, completes the panoply of means of protection against free radicals. These antioxidant enzymes, such as peroxide dismutase, catalase and glutathione peroxidase, are preventive in that they act on the species involved in initiating the free radical reaction chain; whereas smaller antioxidant molecules such as ascorbate, tocopherol, ubiquinone, urea and GSH are capable of trapping oxidising radicals directly and are therefore antioxidants which "break" the free radical chain (BUETTNER, 1993). Generally speaking, the respective actions of the various antioxidant enzymes are not of the same order of importance (HARRIS, 1992).

Numerous studies have reported an increase in oxidative stress during obesity, due both to an increase in the production of oxygen radicals and a reduction in antioxidant defence capacity through a drop in antioxidant enzyme activity and antioxidant vitamin levels (Furukawa et al., 2004). Obesity is considered to be a chronic inflammatory state of low intensity, with inflammation accompanied by an increase in oxidative stress in fat cells, favouring the establishment of insulin resistance (ZARROUKI, 2007). Systemic oxidative stress increases

with the degree of obesity. Indeed, increases in TBARs in plasma are correlated with body mass index and hip circumference (OLUSI, 2002).
H_2O_2 production is increased only in the white adipose tissue of obese KKAy mice and not in other tissues (liver, muscle, aorta) suggesting that adipose tissue is the main site of ROS production (FURUKAWA et al., 2004). In addition, induction of obesity in mice on a hyperlipidic diet significantly increases circulating concentrations of lipid peroxidation markers and induces oxidation of plasma albumin. The concentration of carbonylated proteins in adipose tissue is 2 to 3 times higher in obese mice than in controls (GRIMSRUD et al., 2007).
Numerous studies have highlighted the involvement of oxidative stress in ageing (ROMANO and SERVIDDIO, 2010). It has been widely described that, with age, oxidative stress induces molecular damage to lipids, proteins and nucleic acids in different tissues of various species (BOKOV et al., 2004). Ageing is characterised by a gradual decline in biological functions caused by a progressive dysfunction of the various cellular systems for repairing and maintaining homeostasis. As a result, irreversible damage accumulates in lipids, proteins and nucleic acids (HOLLIDAY, 2006; RATTAN, 2008). However, there is no single theory capable of explaining the causes and mechanisms of all aspects of ageing. Nevertheless, the most popular is the "free radical" or "oxidative stress" theory (KREGEL , 2007) devised by HARMAN (HARMAN, 1956). It states that reactive oxygen species (ROS) cause ageing by damaging DNA and oxidising proteins and lipids (LONDOÑO-VALLEJO, 2009). An increase in biological markers of oxidative stress such as malonic dialdehyde (MDA) has been observed during ageing in many species (LANE, 2003; DELATTRE et al., 2005).
Numerous studies have been carried out on human plasma and erythrocytes in order understand the evolution of antioxidant enzymatic activities as a function of age. In this respect, erythrocytes are a suitable cellular model, as they are exposed to continuous oxidative stress due in particular to the production of oxygenated free radicals generated by auto-oxidation of haemoglobin. Enzymes such as superoxide dismutase (Cu,Zn-SOD), glutathione peroxidase, glutathione reductase and glutathione S-transferase form an important defence network against intra-erythrocytic 'oxidative stress'. A study by CEBALLOS-PICOT et al (1992) carried out on the erythrocytes of 167 subjects aged between one month and sixty-three years showed a negative correlation between age and the activities of SOD, glutathione S-transferase and glutathione reductase. However, the correlation was positive for glutathione peroxidase (CEBALLOS-PICO et al., 1992). These results do not corroborate those obtained in other studies. For example, in a population of 239 subjects aged between sixty-five and ninety, BERR et al (1993) did not observe any significant age-related decrease in either SOD or glutathione reductase. The only significant correlation found in this study concerns plasma selenium, which is reduced. Finally, ARTHUR et al (1992), in a study of 1,836 subjects aged between four and ninety years, showed that erythrocyte SOD activity remains more or less constant up to the age of sixty-five; after this age, this activity decreases moderately. Erythrocyte glutathione peroxidase activity increases up to the age of eighteen, remains stable until the age of sixty-five and then declines, but again only moderately. Based on the results of these three studies, it is difficult to form a precise opinion on the evolution of the activity of erythrocyte antioxidant enzymes during ageing. Other investigations into antioxidant vitamins (vit E, vit C) have led to identical conclusions.

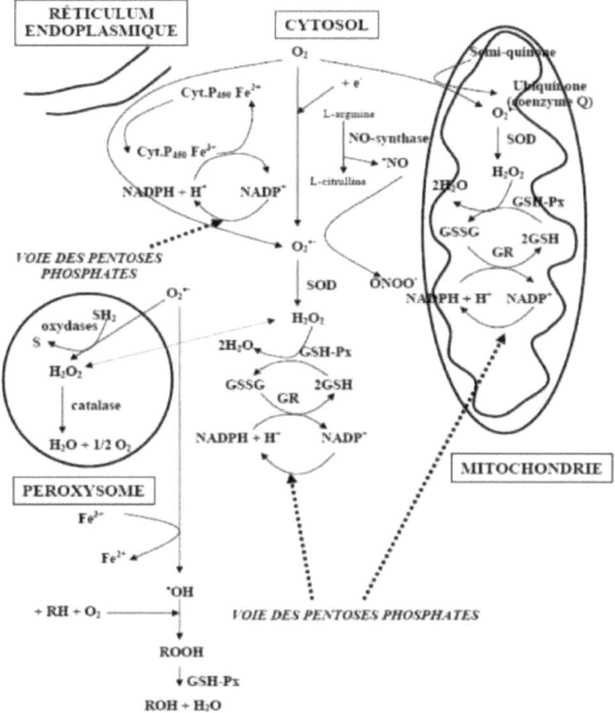

Figure 4: Main cellular sources of free radicals (BONNEFOND-ROUSSELOT et al., 2003).

RH: polyunsaturated fatty acid, ROOH: lipid hydroperoxide, SH2: reducing substrate, S: oxidised substrate, SOD: superoxide dismutase, GSH-Px: glutathione peroxidase, GR: glutathione reductase, GSH: reduced glutathione, GSSG: oxidised glutathione.

III. Effect of linseed oil on obesity and ageing

The role of diet has evolved, with foods increasingly being called upon to provide physiological benefits in terms of disease management and prevention (MAZZA and OOMMAH, 2000). In recent years, there has been growing interest in polyunsaturated fatty acids (PUFAs) and their potential beneficial role in health.

Linseed is one of the richest oils in n-3 PUFAs. Linseed oil is a vegetable oil extracted from the seeds an annual herbaceous member of the *Linum* genus in the Linaceae family: *Linum usitatissimum* (also known by the common name linseed). It is one of the oldest commercial oils, and the processed linseed oil solvent has been used for centuries as a drying oil for paint and varnish. The crude oil is used as an astringent in fungicidal and insecticidal lotions and has shown moderate insect repellent properties (KAITHWAS and MAJUMDAR, 2010).

Flaxseed has been consumed for centuries for its great flavour and range of nutritional benefits revealed by scientific research. Studies suggest that flaxseed in the diet may benefit people with certain types of breast cancer and prostate cancer (LU et al., 2005). Flaxseed can

also reduce the severity of diabetes by stabilising blood sugar levels (DAHL et al., 2005). Seeds can lower cholesterol levels (CUNNANE et al., 1993; PAN et al., 2009).
The benefits of flaxseed oil can be linked to the potential of α-linolenic acid (ALA). According to Simopoulos and Robinson (1998), linseed oil is one of the most highly unsaturated of all oils. It represents the richest plant source of a-linolenic acid (ALA; 5062% flaxseed oil, or (22% whole flaxseed) (PAN et al., 2009). In addition, flaxseed oil contains 12.7% linoleic acid, giving it the highest n-3/n-6 ratio of all plant sources (TZANG et al., 2009).
PUFAs are classified into 4 main families (n-7, n-9, n-6, n-3). The first two families are known as non-essential because their respective precursors, palmitoleic acid and oleic acid, can be synthesised by the body. The n-6 and n-3 are essential because their precursors must be provided by the diet.
α-Linolenic acid (ALA) is the precursor the ω-3 family and linoleic acid (LA) is the precursor of the ω-6 family. The main long-chain GAs derived from these precursors by a process of desaturation and elongation are arachidonic acid (20:4 n6) for the ω-6 family and eicosapentaenoic acid (EPA, 20:5 n3) and docosahexaenoic acid (DHA, 22:6 n3) for the ω-3 family (GIBSON et al., 2011) (Figure 5).

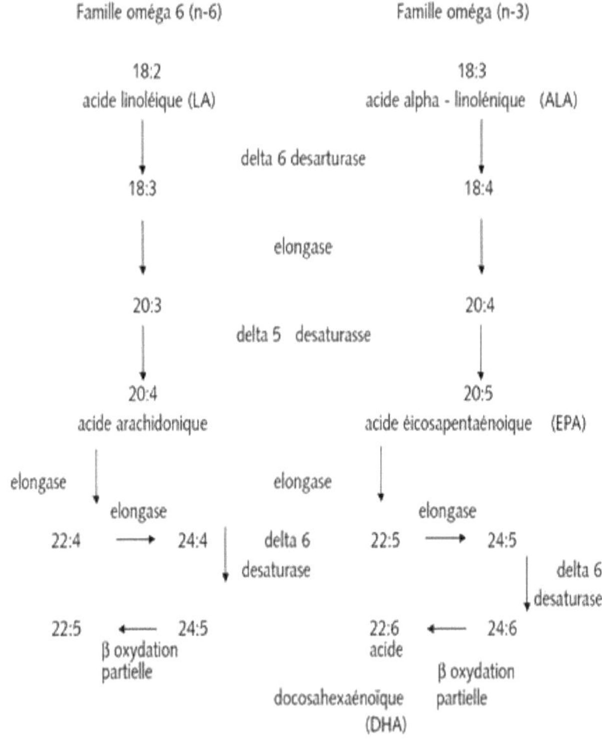

Figure 5: **Biosynthesis of polyunsaturated fatty acids and their derivatives** (AFSSA, 2001).

Several authors have demonstrated the benefits of n-3 PUFAs on dyslipidemia and oxidative stress (YILMAZ et al 2002; AILHAUD et GUESNET, 2003; MERZOUK et KHAN, 2003). DE LORGERIL et al (1994) showed that a Cretan diet rich in alpha-linolenic acid reduced the rate of myocardial infarction by more than 70%. The physiological effects of n-3 PUFAs occur at several levels. Firstly, they play a structural role in cell membranes, as they are the natural constituents of membrane phospholipids. They regulate the activity of enzymes, transporters and receptors in biological membranes, either directly or indirectly, by modifying their physico-chemical properties. The second level involves the process of communication between cells, as these PUFAs are essential precursors of the eicosanoid synthesis pathway, which regulates functions as diverse as reproduction, cardiac physiology, blood coagulation, inflammation and the functioning of endocrine and exocrine glands, and groups together various types of molecules such as prostaglandins. Finally, PUFAs interact in the nucleus of adipose and liver cells, regulating the expression of genes involved in their transport in blood in the form of lipoproteins, and their metabolism via nucleic receptors (DURAND et al., 2002). Studies have shown that n-3 fatty acids contained in fish oil are beneficial in the prevention and treatment of patients suffering from cardiovascular disease (XIN et al., 2013). They have several sites of action and therefore several effects, such as the prevention of arrhythmias, anti-coagulant and anti-platelet action, modulation of cell growth in the arterial wall, improvement in vascular haemodynamics, regulation of blood pressure and lipid-lowering action. With regard to this last property, it has been observed that these fatty acids reduce plasma triglyceride levels and LDL cholesterol synthesis, while increasing HDL cholesterol and the activity of the enzyme lecithin cholesterol acyl transferase (LCAT) (FUMERON et al., 1991).

SIRIWARDHANA et al (2013) have shown the benefits of n-3 PUFAs on obesity by reducing fat mass, stimulating lipolysis and fatty acid oxidation and inhibiting lipogenesis and the production of pro-inflammatory adipo-cytokines (Figure 6).

Furthermore, in the presence of ROS, n-6 PUFAs oxidise easily to form lipid peroxides. In addition, the biosynthesis of these fatty acids leads to the formation of pro-inflammatory prostaglandins, whereas n-3 PUFAs (alphalinoleic and docosahexaenoic acids) are less easily peroxidised and their biosynthesis gives rise to anti-inflammatory prostaglandins. n-3 PUFAs thus stimulate

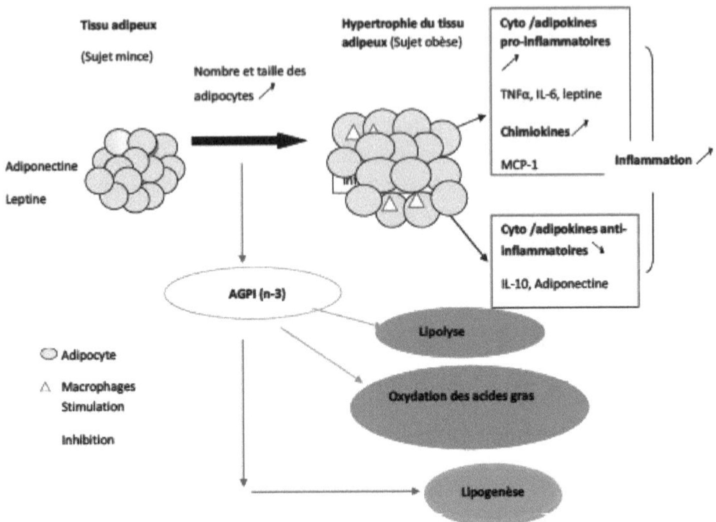

Figure 6: Effect of omega-3 on adipose tissue dysfunction associated with obesity (SIRIWARDHANA et al., 2013)
IL-6: interleukin 6
IL-10; interleukin 6
TNFa: tumour necrosis factor
MCP-1: cytokine called (Monocyte chemoattractant protein 1)

the activity of antioxidant enzymes, which makes a major contribution to improving the impaired oxidant/antioxidant status of obese people (SARSILMAZ et al., 2003).

Supplementation with n-3 PUFAs improves glycaemic control and increases insulin sensitivity in obese people. n-3 PUFAs are rapidly metabolised and therefore less stored, hence their favourable effects on the energy balance in obese people (STORLIEN et al., 1998).

The beneficial effects of n-3 PUFAs on oxidative stress are becoming increasingly evident. They modulate the activities of antioxidant enzymes and increase the effectiveness of the body's antioxidant defence system (DEMOZ et al., 1992; VENKATRAMAN et al., 1994).

MAKNI et al (2008) evaluated the lipid-lowering and hepatoprotective effects of a mixture of flaxseed and pumpkin seeds rich in n-3 and n-6 PUFAs in hypercholesterolaemic rats. ABDEL-MONEIM et al (2010), in a study on male albino rats, evaluated the effect of linseed oil on oxidative stress induced by lead acetate and its toxicity in the liver, and showed that linseed oil significantly reduces liver damage and inflammation and, obviously, acts as a hepatoprotective agent.

CHAPTER 2

MATERIALS AND METHODS

I. Experimental protocol

I.1. Animals and diets

The study was carried out on Wistar rats bred at the animal house in the Biology Department, Faculty of Natural and Life Sciences and Earth and Universal Sciences, Aboubekr BELKAID-TLEMCEN University. The animals have free access to water and food.

Male rats aged 18 months and weighing between 320 and 350 g were separated into six groups and fed six different diets:

- **A control batch (C)**: consisting of 10 rats eating the standard commercial diet.
- **A flax 2.5% control batch (CL2.5%):** consisting of 10 rats consuming the standard diet enriched with 2.5% flaxseed oil.
- **A 5% flax control batch (CL5%):** consisting of 10 rats consuming the standard diet enriched with 5% flaxseed oil.
- **An obese batch (CAF):** made up of 10 rats consuming the cafeteria diet, consisting of 30 g of standard diet and 30 g of a mixture of pâté-dry biscuits-cheese-chips-chocolate-peanuts in the proportions 2:2:2:1:1:1 according to the DARIMONT et al. protocol (2004).
- **An obese flax 2.5% batch (CAFL2.5%):** consisting of 10 rats consuming the cafeteria diet enriched with 5% flaxseed oil.
- **An obese flax 5% batch (CAFL5%):** consisting of 10 rats consuming the cafeteria diet enriched with 5% flaxseed oil.

The composition of the different diets is given in Table 1.

The rats were fed the diet for two months. Body weight and food intake were recorded daily.

I.2. Sacrifices and blood sampling

After 2 months on the diet, the rats were anaesthetised with sodium pentobarbital (60 mg/kg body weight) and sacrificed after 12 h of fasting. Blood was collected by puncture of the abdominal aorta. Some of the blood was collected in EDTA tubes and some in dry tubes. Samples collected in EDTA tubes are centrifuged at 3,000 rpm for 15 minutes. Plasma is collected for the determination of oxidative stress parameters (vitamin C, MDA, hydroperoxides, protein carbonyls, conjugated dienes).

Table 1: Composition of the diets consumed by the rats

Composition (% by weight)	Diet					
	C	CL2.5% (%)	CL5%	CAF	CAFL2.5	CAFL5%
Proteins	19	19	19	21	21	21
Carbohydrates	56	56	56	31	31	31
Lipids	3,5	3,5	3,5	30	30	30
Sunflower oil	5	2,5	0	5	2,5	0

Linseed oil	0	2,5	5	0	2,5	5
Fibres	8	8	8	4	4	4
Humidity	5	5	5	4	4	4
Minerals	1	1	1	1	1	1
Vitamins	4	4	4	4	4	4
Composition (% energy)						
Proteins	20	20	20	16	16	16
Carbohydrates	60	60	60	24	24	24
Lipids	20	20	20	60	60	60
Energy (Kcal/100g)	330	330	330	420	420	420
Fatty acids:						
AGS	27	20	22	42	32	22
AGMI	24	18	15	30	24	26
C18:2n-6	45	36	28	27	20	23
C18:3n-3	3	25	34	1	23	28
C20:4n-6	1	1	1	0	1	1

The composition of the diets was determined at the Natural Products Laboratory, SNVTU Faculty, University of Tlemcen. Fatty acid composition was determined by HPLC at the UPRES lipids laboratory, Gabriel Faculty of Science, University of Burgundy Dijon, France. C: rats on control diet; CL2.5%: rats on control diet enriched with 2.5% linseed oil; CL5%: rats on control diet enriched with 5% linseed oil; CAF: rats on cafeteria diet; CAFL2.5%: rats on cafeteria diet enriched with 2.5% linseed oil; CAFL5%: rats on cafeteria diet enriched with 5% linseed oil.

The remaining erythrocytes were washed three consecutive times with physiological water, then lysed by the addition of ice-cold distilled water (1/3, v/3v), and incubated for 15 min on ice. Cellular debris was removed by centrifugation at 5000 rpm for 5 min. The lysate was then recovered for assay of the erythrocyte antioxidant enzyme catalase and reduced glutathione.

After coagulation of the blood taken from dry tubes, and centrifugation at 3000 rpm for 15 min, the serum is recovered and stored with a 0.2% NaN3 and 10% Na2 EDTA solution, at a rate of 10 µľ/ml, at -20°C with a view to assaying the various parameters of lipoprotein metabolism and determining fatty acid composition.

Glucose, vitamin C and LCAT are measured on the same day as the sample.

I.3. Organ retrieval

After blood sampling, the liver, gastronemic muscle, visceral adipose tissue and part of the intestine were carefully removed, rinsed with 0.9% NaCl and weighed. An aliquot of the various organs was immediately ground in PBS buffer, pH 7.4. The homogenate obtained was used to determine the various parameters of oxidant/antioxidant status. Another aliquot of adipose tissue was homogenised in the same grinding buffer supplemented with 20 mg/ml leupeptin, 2 mg/ml antipain and 1 mg/ml pepstatin (leupeptin, antipain and pepstatin are inhibitors of proteases and therefore of intracellular proteolysis) according to the protocol of KABBAJ et al. (2003). The homogenate obtained was used to determine the activity of the LHS enzyme. For LPL activity, target organ homogenates are prepared in a 0.9% (w/v) NaCl solution containing heparin (Sigma, St. Louis, MO, U.S.A.), according to MATH et al. (1991).

The organ remains are stored at -80°C for lipid and protein assays and to determine their fatty acid composition.

II. Biochemical analysis

II.1. . Determination of glucose, urea and creatinine levels

❖ Serum glucose is determined enzymatically and colorimetrically in the presence of glucose oxidase (GOD). Glucose is oxidised to gluconic acid and hydrogen peroxide. In the presence of peroxidase and phenol, hydrogen peroxide oxidises a colourless chromogen (4-aminoantipyrine) to a red colour with a quinoneimine structure. The colour obtained is proportional to the concentration of glucose in the sample. The reading is taken at a wavelength of 505 nm (Prochima kit).

❖ Serum creatinine is determined by the Jaffé method, based on the reaction of picric acid with creatinine in a basic medium, forming a yellow-orange coloured complex. The intensity of the colouration is measured at a wavelength of 530 nm (Prochima kit).

❖ Serum urea is determined using a colorimetric method based on the use of diacetylmonooxine and $Fe^{(+3)}$ ions (Prochima Kit). In the presence of $Fe^{(+3)}$ ions and a reducing agent, urea reacts with diacetylmonooxine to give a pink-coloured complex. The colouration obtained is proportional to the quantity of urea present in the sample. The reading is taken at a wavelength of 525 nm.

II.2. . Determination of total protein :

Total protein was determined on lipoprotein fractions and organs (after grinding as described above) by the method of LOWRY et al (1951) using bovine serum albumin as the standard (Sigma Chemical Company, St Louis, MO, USA). In an alkaline medium, the complex formed by Cu^{2+} ions and the tyrosine and tryptophan groups of the proteins is reduced by Folin's reagent. The blue colour developed is proportional to the amount of protein in the sample. The reading is taken at a wavelength of 689 nm.

Total protein in serum is determined using the biuret reagent (Prochima kit).

II.3. . Determination of lipid parameters in serum, lipoproteins and organs

II.3.1. . Separation of lipoproteins

Total lipoproteins are isolated from serum by precipitation using the method of BURSTEIN et al (1970, 1989). At neutral pH, polyanions, in the presence of divalent cations, can form insoluble complexes with lipoproteins (lipopolyanion-cations), so precipitation takes place thanks to the polyanions which combine with the lipids in the lipoproteins. Generally, the polyanions used are sulphates (SO^{3-}), polysaccharides (heparin) and phosphotungstic acid, while the cations are Ca^{2+}, Mn^{2+} and Mg^{2+}. By using the same precipitation reagent at different concentrations, it is possible to selectively precipitate lipoprotein fractions; and so, at increasingly higher concentrations, this reagent allows the separation from serum, firstly of VLDL, then of LDL and lastly of HDL. This principle is similar to that of density gradient ultracentrifugation of lipoproteins. When the concentration varies, the density of the medium also varies, allowing selective precipitation . The lipoproteins, precipitated by phosphotungstic acid and MgCl2 at different concentrations, are then solubilised using a solubilisation solution containing trisodium citrate buffer and NaCl.

II.3.2. . Determination of cholesterol and triglyceride levels

Total cholesterol and triglycerides were determined using enzymatic methods (Prochima kit) on total serum, the various lipoprotein fractions and organ homogenates (prepared by grinding an aliquot in phosphate/EDTA buffer, pH 7.2, adding 1% sodium lauryl sulphate (SDS) (1/1, v/v), and centrifuging at 3000 rpm for 10 min).

II.3.3. . Extraction of lipids from serum and organs and determination of fatty acid composition

Lipid extraction was carried out from serum using a mixture of methanol/chloroform/NaCl (2M) (1/1/0.9; v/v/v) for 0.5 ml of sample, according to the method of BLIGH and DYER (1959). Organ lipids (liver, gastrocnemius muscle, adipose tissue) were extracted using the method of FOLCH et al (1957) after grinding an aliquot of the organ (300 mg) in 1 ml of NaCl (2M) using the ultraturax (Bioblockscientific, III Kirch, France).

A specified amount (30 μï) of internal standard (heptadecanoic acid 17 :0, $C_{17}H_{34}O2$ diluted in benzene to 2mg/ml) is added.

After extraction, the fatty acids were saponified with 1ml of 0.5N methalonic NaOH, vortexed and heated to 80°C for 15 min. The reaction was stopped by thermal shock by placing the tubes in ice. The fatty acids were then methylated by adding 2 ml of BF3 methanol (14% Bromotrifluoromethanol) (SLOVER and LANZA, 1979). After vortex mixing, the tubes were closed under nitrogen and baked at 80°C for 20 min. The reaction was stopped by heat shock. 2 ml of saturated NaCL (35%) and 2 ml of hexane were then added to the tubes. After vortexing, two phases were formed. The upper phase was taken injection into the chromatograph.

The analysis was carried out by GC (gas chromatography; Becker instruments, Downersgrove, IL); the capillary column (Applied Sciences Labs, State College, PA) was made of Pyrex, 50 m long and 0.3 mm internal diameter, filled with 20 m carbowax (Spiral-RD, Couternon, France). The chromatograph is equipped a ROS injector and a flame ionisation detector connected to an Enica 21 integrator-calculator (DELSI instruments, Suresnes, France). Fatty acids were identified by comparing their retention times with those of fatty acid standards (Nucheck-prep, Elysiam, MN, USA). The area of the fatty acid peaks is proportional to their quantity and is calculated using an integrator.

II.4. . Determination of lecithin cholesterol acyltransferase (LCAT, EC 2.3.1.43) activity

LCAT activity is determined on fresh serum by estimating the conversion of serum cholesterol to esterified cholesterol using the method of ALBERS et al (1986). Esterified cholesterol is measured before and after incubation of fresh serum at 37°C for 1 hour. Esterified cholesterol is measured after precipitation of free cholesterol by digitonin (BAUER, 1982). The increase in esterified cholesterol levels corresponds to the enzymatic activity of LCAT, which is expressed in nmoles of esterified cholesterol/h/ml of serum.

II.5. . Tissue lipase assay

II.5.1. . Determination of LPL enzyme activity (LPL, EC 3.1.1.34)

Lipase activity is determined from the level TG hydrolysis of a synthetic substrate by measuring the quantity of fatty acids released by titrimetry using the PH - STAT technique (TAYLOR, 1985; TIETZ et al., 1989). An emulsion of olive oil (20 mL) and gum arabic (16.5 g) solubilised in $H2O$ (165 mL) was prepared by sonication (3 times 45 minutes). The synthetic substrate contains 2.4 mL of emulsion, 300μl of bovine serum albumin solution (4% in tris/HCl buffer) and 300 μl of human serum heated to 56°C. 100 uL of synthetic substrate is incubated with 100 uL of supernatant (enzyme source) in 3 mL of 100 mmol/L NaCl, 5 mmol/L CaCl2 buffer; PH 8, at room temperature with shaking for 5 min. After incubation, the pH of the medium (which had become acidic due to the release of GLA) was returned to its initial value by adding 0.05 mol/L NaOH. The volume of NaOH added is then noted and, after conversion, corresponds to the number of fatty acids released (mol).

A lipase unit is the quantity of enzyme that enables the release of one micromole of fatty acid in one minute.

II.5.2. . Determination of hormone-sensitive lipase enzyme activity (LHS; EC 3.1.1.3)

Lipolytic activity was measured quantitatively using the method described by KABBAJ et al (2003). This activity is measured with *p-nitrophenyl butyrate* ester (PNPB), hydrolysed in the presence of lipase to *p-nitrophenol* and butyric acid. The release of *p-nitrophenol* results in the appearance of a yellow coloration detected at 400 nm. The adipose tissue homogenates were incubated with PNPB in buffer (0.1 M NaH2PO4, pH 7.25, 0.9% NaCl, 1 mMdithiothreitol) at 37°C for 10 minutes. The reaction was stopped by adding a mixture of methanol/chloroform/heptane (10/9/7). After centrifugation at 800g for 20 minutes, the solutions were incubated for 3 minutes at 42°C. The absorbance read at 400 nm was used to calculate the concentration using a molar extinction coefficient of 12.75 10^3 M^{-1} cm^{-1} for *p-nitrophenol.*

An enzyme unit is the amount of enzyme capable of releasing one μmole of *p*- nitrophenol per minute per mg of protein.

III. Determining oxidant/antioxidant status

III.1. 1. Determination of vitamin C

Plasma vitamin C is determined according to the method of ROE and KUETHER (1943) using the staining reagent Dinitrophenylhydrazine-Thiourea-Copper (DTC) and a standard range ascorbic acid.

After precipitation of the plasma proteins with trichloroacetic acid (10%) and centrifugation, an aliquot of the supernatant was mixed with the DTC reagent (9 N sulphuric acid, 3% 2,4-dinitrophenylhydrazine, 0.4% thiourea and 0.05% copper sulphate). The mixture was incubated for 3 hours at 37°C. The reaction was stopped by 65% (v/v) sulphuric acid, and the absorbance was read at 520 nm. The intensity of the colour obtained is proportional to the concentration of vitamin C present in the sample. The concentration is determined from a standard curve obtained using an ascorbic acid solution.

III.2. 2. Determination of carbonylated proteins

Carbonylated plasma or tissue proteins (markers of protein oxidation) are measured by the 2,4- dinitrophenylhydrazine reaction using the method of LEVINE et al. (1990). Plasma or organ homogenate was incubated for 1 h at room temperature in the presence of dinitrophenylhydrazine (DNPH; prepared in HCL) or with HCL alone for the blank. The proteins were then precipitated trichloroacetic acid (TCA) and washed 3 times with ethanol: ethylacetate 1:1 (v/v) and 3 times with TCA.

The pellet is solubilised in a guanidine solution.

Readings were taken at 350 and 375nm. The concentration of carbonyl groups is calculated using an extinction (E = 21.5 $mmol^{-1}$. L. cm^{-1}).

III.3. 3. Determination of hydroperoxide

Plasma hydroperoxides are measured by oxidation ferric ions using xylenol orange (Rockford, IL, USA) in conjugation with the specific reducing ROOH of triphenylphosphine (TPP), according to the method of NOUROOZ-ZADEH et al. (1996). This method is based on rapid peroxidation transforming Fe2+ into Fe3+ in an acid medium. Fe3+ ions in the presence of orange xylenol [(O-cresolsulfonphthalein-3',3''-bis (methyliminodiacetic acid sodium)], form an Fe^{3+}-xylenol orange complex. The absorbance of this coloured complex is measured at 560 nm. The level plasma hydroperoxides corresponds to the difference between the absorbance of the sample and the absorbance of the blank.

III. 4. Determination of malondialdehyde (MDA)

Malondialdehyde (MDA) is the most widely used marker of lipid peroxidation. The assay is based on the method of Draper et Adley al (1996), using a hot acid treatment with thiobarbituric acid (TBA). Plasma or organ homogenate was incubated for 20 minutes at 100°C with TBA and trichloroacetic acid (TCA). After incubation, cooling and centrifugation at 4,000 rpm for 10 minutes, the reading is taken on the supernatant containing the MDA. TBA reacts with aldehydes to form a chromogenic condensation product consisting of 2 molecules of TBA and one molecule of MDA, which absorbs at 532 nm. The concentration of plasma or tissue MDA is calculated using an MDA standard curve or the extinction coefficient of the MDA-TBA complex ($E= 1.56*10^{5} mol^{-1}.L.cm^{-1}$ at 532 nm).

III.5. 5. In vitro oxidation of serum lipoproteins, induced by metals (copper), is determined by monitoring the formation of conjugated dienes over time using the method of ESTERBAUER et al. (1989). The dienes are considered to be the primary products of lipid oxidation and have an ultraviolet absorption at 234 nm. The addition of $CuSO_4$ (IOOμM) to serum causes oxidation of serum lipoproteins which results in a progressive increase in optical density at 234 nm, after a latent phase. This increase in absorbance marks the increasing formation of conjugated dienes, the concentration of which is estimated using the extinction coefficient ($E = 29.50 mmol^{-1}. L.cm^{-1}$; at 234 nm). Variations in the absorbance of the conjugated dienes as a function of time are used to plot the kinetic curve, in which three consecutive phases are determined: latent phase, propagation phase and decomposition phase. Several markers of in vitro oxidation of plasma lipoproteins are determined from this kinetic curve:

- t (lag) min corresponds to the duration of the lag phase and marks the start of the increase in optical density compared with the initial value. The t (lag) is used to estimate the resistance of lipoproteins to oxidation in vitro.
- Initial rate of conjugated dienes (μmol/l)
- Maximum rate of conjugated dienes (μmol/l)
- t (max) min is the time required to maximum oxidation. It marks the end of the propagation phase and the start of the decomposition phase. It is calculated on the kinetic curve by projecting the maximum optical density value onto the X axis (time expressed in minutes).

The rate lipoprotein oxidation is calculated by :

(Maximum conjugated diene rate - Initial conjugated diene rate)/t(max)- t(lag).

III.6. 6. Determination of catalase activity (CAT; EC 1.11.1.6)

This enzymatic activity is measured by spectrophotometric analysis of the rate of hydrogen decomposition (AEBI, 1974). In the presence of catalase, the decomposition of hydrogen leads to a decrease in the absorption of the H_2O_2 solution as a function of time. The reaction medium contains erythrocyte lysate diluted 1:500, H_2O_2 and phosphate buffer (50 mmol/l, pH 7.0). After incubation, the staining reagent, titanium oxide sulphate ($TiOSO_4$) (prepared in 2N H_2SO_4) was added. Readings were taken at 420 nm. The concentrations of the remaining H_2O_2 are determined using a standard H_2O_2 range with phosphate buffer and $TiOSO_4$ reagent to obtain concentrations of 0.5 to 2 mmol/l in the reaction medium. The calculation one unit of enzyme activity is :

$A = \log_{A1} - \log_{A2}$.

A1 is the starting H2O2 concentration

A2 is the concentration of H2O2 after incubation

Specific activity is expressed in U/g Hb or U/ml.

III.7. 7. Determination of reduced Glutathione (GSH):

Reduced glutathione (GSH) is determined by the colorimetric method using Ellman's reagent (DTNB) (ELLMAN, 1959). The reaction consists of cutting the 5,5dithiodis-2-nitrobenzoi'que acid molecule (DTNB) with GSH, which releases thionitrobenzoic acid (TNB) according to the following reaction:

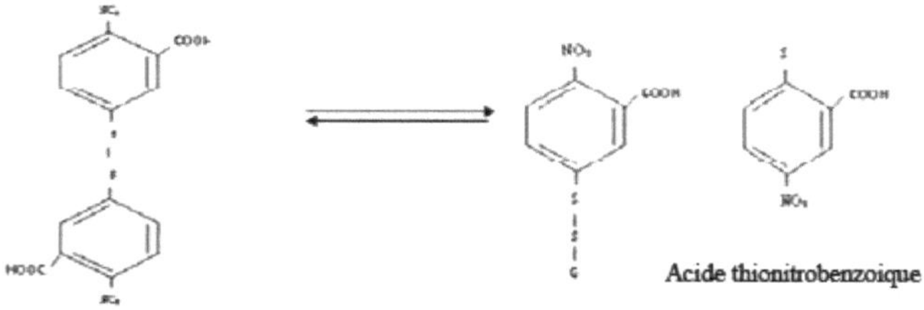

Acide thionitrobenzoique (TNB)

DTNB

Thionitrobenzoic acid (TNB) at alkaline pH (8-9) exhibited an absorbance at 412 min with an extinction coefficient equal to 13.6 $mM^{-1}.cm^{-1}$.

IV. Statistical analysis

The results are presented as the mean ± standard error. After analysis of variance, the means between the three groups of control rats and between the three groups of obese rats were compared using a one-factor ANOVA test. This analysis was completed using Tukey's test to classify and compare the means in pairs. Means indicated by different letters (a, b, c, d) are significantly different ($P < 0.05$).

CHAPTER 3

RESULTS AND INTERPRETATION

I. Body weight, food consumption and daily energy intake in different batches of rats (Figure 7, Table A1 in appendices)

I.I. Body weight in control and experimental rats (Figure 7, Table A1 in appendices)

At the start of the experiment, the rats used in this study were of uniform weight.
(± 20 g) and fed for two months on the different diets: standard (C), standard enriched with 2.5% linseed oil (CL2.5%), standard enriched with 5% linseed oil (CL2.5%) cafeteria (CAF), cafeteria enriched with linseed oil (CAFL2.5%) and (CAFL5%). At the end of the experiment, rats fed the cafeteria diet (CAF) enriched or not with linseed oil showed a significant increase in weight compared with their respective controls. However, the weight gain observed in rats fed the cafeteria diet enriched with linseed oil (CAFL) was significantly lower than in rats fed the cafeteria diet alone (CAF). Flaxseed oil reduced the weight gain induced by the cafeteria diet. The control diet (C) and the cafeteria diet supplemented with 5% linseed oil did not induce any variation in body weight compared with the diet supplemented with 2.5% linseed oil. Overall, the cafeteria diet *resulted in* an increase in body weight in aged rats after the first week of the diet, with this weight variation becoming more significant after two months on the diet. Supplementation of the diet with linseed oil resulted in a slight weight loss in the controls, but this was significant in the obese rats (cafeteria).
The adiposity index (Table 2) was significantly higher in obese rats than in control rats. However, linseed oil significantly reduced adiposity in rats on the cafeteria diet.

I.2 Variation in food consumption (Figure 7, Table A1 in appendices)

Over the experimental period (60 days), food consumption expressed as (g/d/rat) in control and experimental rats varied significantly between the different batches. A significant increase in food intake was noted in rats fed the cafeteria diet compared with rats fed the standard diet.
However, rats fed the cafeteria diet enriched with 2.5% and 5% CAFL linseed oil ingested significantly more food than rats fed the standard diet enriched with 2.5% and 5% CL linseed oil, respectively, and more than rats fed the cafeteria diet alone.

I.3. Daily energy intake in control and experimental rats (Figure 7, Table A1 in appendices).

The energy intake (expressed as Kcal/J/rat) of obese rats fed the cafeteria diet enriched or not with linseed oil (CAF, CAFL2.5%, CAFL5%) increased significantly compared with their respective controls (C,CL2.5%,CL5%).
However, there was a significant reduction in energy intake in the batches receiving the control and cafeteria diet supplemented with linseed oil compared with the batches receiving the control and cafeteria diet not supplemented with linseed oil.

II. Blood biochemical parameters

ILl.serum glucose, urea and creatinine levels in control and experimental rats (Figure 8, table A2 in appendices)

Elderly obese rats fed a cafeteria diet supplemented or not with linseed oil (CAF, CAFL2.5%, CAFL5%) showed a significant increase in serum glucose levels compared with their respective controls (C, CL2.5%, CL5%).

Flaxseed oil significantly reduced blood sugar levels in (CAFL2.5%) and (CAFL5%) rats compared with (CAF) rats. The reduction was more pronounced for (CAFL5%) than (CAFL2.5%) rats, at around (2.7%).

However, the control rats (CL2.5% and CL5%) showed no variation in serum glucose levels compared (C).

Rats receiving the cafeteria diet (CAF) showed an increase in serum urea and creatinine levels compared with the other batches.

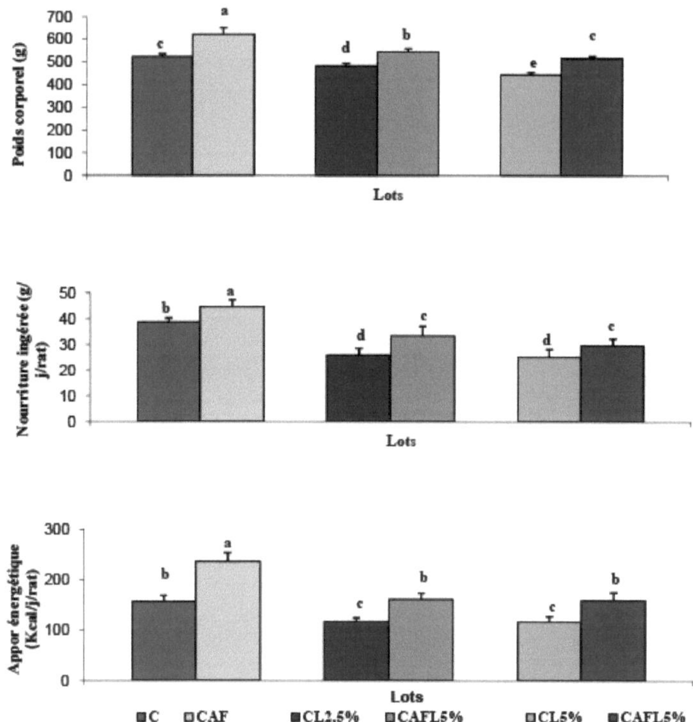

Figure 7: Body weight, food intake and energy intake in different batches of rats

Each value represents the mean ± SE, n=10.C: control rats fed the standard diet; CAF: obese rats fed the cafeteria diet; CL2.5%: control rats fed the standard diet enriched with 2.5% linseed oil; CAFL2.5%: obese rats fed the cafeteria diet enriched with 2.5% linseed oil; CL5%: control rats fed the standard diet enriched with 5% linseed oil; CAFL5%: obese rats fed the cafeteria diet enriched with 5% linseed oil. After verifying the normal distribution of the variables (Shapiro-Wilk test), the means of the six groups of rats were compared using a one-factor ANOVA test. This analysis was completed by the Tukey test in order to classify and compare the means in pairs. Means indicated by different letters (a, b, c) are significantly different ($p<0.05$).

Table 2: Body fat in different batches of rats

Lots Parameter	Standard rats **(C)**	Rats cafeteria **(CAF)**	Standard flax rats2.5% **(CL2.5%)**	Rats cafeteria lin2,5% **(CAFL2,5)**	Standard rats lin5% **(CL5%)**	Rats cafeteria lin5% **(CAFL5%)**	*P* **(ANOVA)**
Body fat index	1,36±0,18^{d}	2,56±0,22 a	1,35±0,10^{d}	2,29±0,28b	1,01±0,12^{e}	1,59±0,14^{c}	0,001

Each value represents the mean ± SE, n=10.C: control rats fed the standard diet; CAF: obese rats fed the cafeteria diet; CL2.5%: control rats fed the standard diet enriched with 2.5% linseed oil; CAFL2.5%: obese rats fed the cafeteria diet enriched with 2.5% linseed oil; CL5%: control rats fed the standard diet enriched with 5% linseed oil; CAFL5%: obese rats fed the cafeteria diet enriched with 5% linseed oil. After verifying the normal distribution of the variables (Shapiro-Wilk test), the means of the six groups of rats were compared using a one-factor ANOVA test. This analysis was completed by the Tukey test in order to classify and compare the means in pairs. Means indicated by different letters (a, b, c) are significantly different ($p<0.05$).

Adiposity index= adipose tissue weight/muscle weight g/g

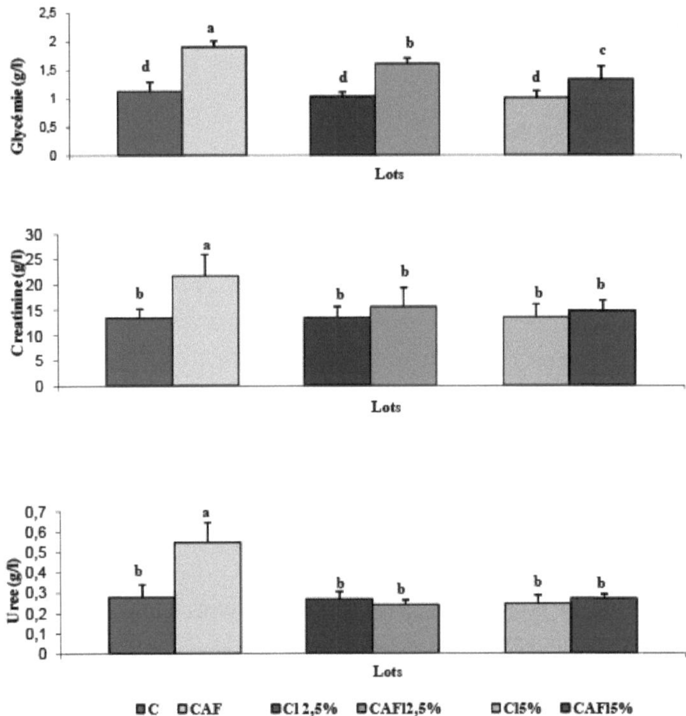

Figure 8: Plasma levels of glycaemia creatinine and urea in different batches of rats

Each value represents the mean ± SD, n=10.C: control rats fed the standard diet; CAF: obese rats fed the cafeteria diet; CL2.5%: control rats fed the standard diet enriched with 2.5% linseed oil; CAFL2.5%: obese rats fed the cafeteria diet enriched with 2.5% linseed oil;

CL5%: control rats fed the standard diet enriched with 5% linseed oil; CAFL5%: obese rats fed the cafeteria diet enriched with 5% linseed oil. After verifying the normal distribution of the variables (Shapiro-Wilk test), the means of the six groups of rats were compared using a one-factor ANOVA test. This analysis was completed by the Tukey test in order to classify and compare the means in pairs. Means indicated by different letters (a, b, c) are significantly different ($p<0.05$).

II.2. Serum and lipoprotein cholesterol levels (mg/dl) in control and experimental rats (Figures 9, Table A3 in appendices).

Serum total cholesterol levels increased significantly in rats (CAF) compared with the other batches. No other variations were noted.

In terms of lipoproteins, there was a significant decrease in HDL cholesterol in the obese rats (CAF, CAFL2.5% and CAFL5%) compared with their respective controls (C, CL2.5%, CL5%), with the decrease being more pronounced in the (CAF) rats. In addition, a significant increase in HDL cholesterol was noted in rats (CL2.5% and CL5%) compared with rats (C).

LDL-cholesterol levels were significantly increased in obese rats on the cafeteria diet (CAF) compared with control rats (C). However, a significant decrease in LDL-cholesterol levels was noted in rats (CAFL2.5% and CAFL5%) compared with rats (CAF) and in rats (CL5%) compared with rats (CAFL5%). LDL-cholesterol levels were also significantly reduced in the 2.5% and 5% lin control rats (Cl2.5% and CL5%) compared with the control rats (C).

Obese rats on the cafeteria diet (CAF) showed a significant increase in VLDL cholesterol levels compared with control rats (C). Administration of linseed oil to the groups of rats on the cafeteria diet (CAFL2.5 and CAFL5%) significantly reduced VLDL cholesterol levels compared with the cafeteria rats (CAF). A significant reduction was also observed in control rats on the standard flax diet (CL2.5% and CL5%) compared with control rats (C).

II.3. Triglyceride levels (mg/dl) in serum and lipoproteins in control and experimental rats (Figure 10, Table A4 in appendices).

Serum triglyceride levels were significantly increased in rats on the cafeteria diet (CAF) compared with control rats (C) and the other batches. The combination of linseed oil and the cafeteria diet significantly reduced the concentration of serum triglycerides in rats (CAFL2.5 and CAFL5%) compared with rats on the cafeteria diet alone (CAF), with the reduction being more pronounced for (CAFL5%) than (CAFL2.5%). There was also a significant reduction in serum triglyceride levels in the flax control rats (CL2.5% and CL5%) compared with the rats on the control diet (C). The reduction was (6%) greater in the (CL5%) rats compared with the (CL2.5%) rats.

With regard to HDL, a significant increase in triglyceride levels was observed in rats on the cafeteria diet (CAF) compared with control rats (C). Flaxseed oil significantly reduced HDL-triglyceride levels on both the cafeteria and standard diets.

A significant increase in LDL-triglyceride levels was observed in rats receiving the cafeteria diet (CAF) compared with their controls (C). Flaxseed oil significantly reduced LDL-triglycerides in rats (CAFL2.5% and CAFL5%) compared with rats on the cafeteria diet alone (CAF). There was also a significant reduction of 4% in rats (CL5%) compared with rats (CL2.5%).

Rats on the cafeteria diet (CAF) showed a significant increase in VLDL-triglyceride levels compared with control rats (C). Administration of linseed oil to the groups of rats receiving the cafeteria diet (CAFL2.5 and CAFL5%) significantly reduced VLDL-triglyceride levels compared with the cafeteria rats (CAF). A significant reduction in VLDL-triglycerides was also observed in rats on the standard flax diet (CL2.5% and CL5%) compared with control rats (C), with a 4.9% reduction in rats (CL5%) compared with rats (CL2.5%).

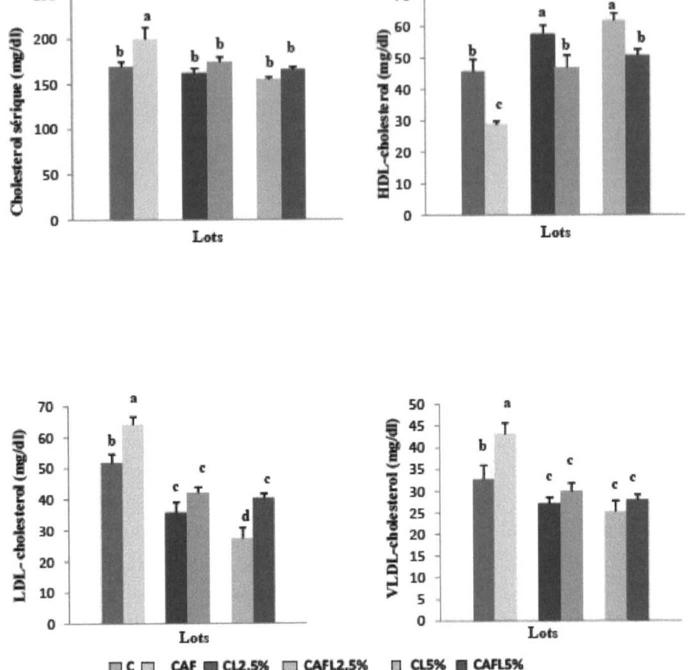

Figure 9: Cholesterol levels (mg/dl) in serum and lipoproteins in different batches of rats
Each value represents the mean ± SD, n=10.C: control rats fed the standard diet; CAF: obese rats fed the cafeteria diet; CL2.5%: control rats fed the standard diet enriched with 2.5% linseed oil; CAFL2.5%: obese rats fed the cafeteria diet enriched with 2.5% linseed oil; CL5%: control rats fed the standard diet enriched with 5% linseed oil; CAFL5%: obese rats fed the cafeteria diet enriched with 5% linseed oil. After verifying the normal distribution of the variables (Shapiro-Wilk test), the means of the six groups of rats were compared using a one-factor ANOVA test. This analysis was completed by the Tukey test in order to classify and compare the means in pairs. Means indicated by different letters (a, b, c) are significantly different ($p<0.05$).

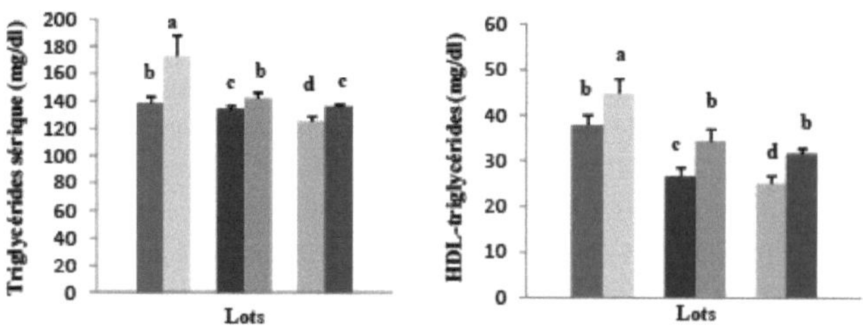

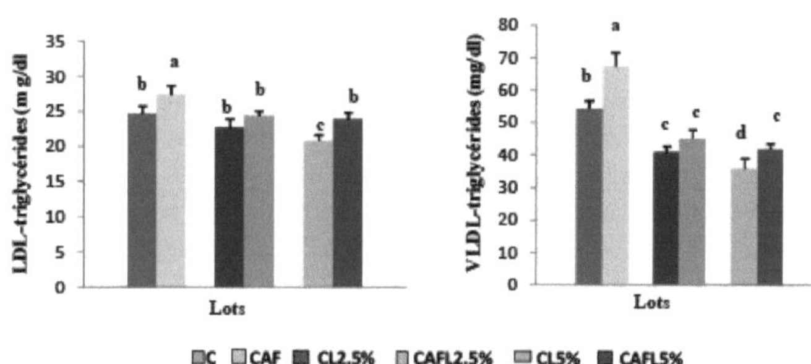

Figure 10: Triglyceride levels (mg/dl) in serum and lipoproteins in different batches of rats

Each value represents the mean ± SD, n=10.C: control rats fed the standard diet; CAF: obese rats fed the cafeteria diet; CL2.5%: control rats fed the standard diet enriched with 2.5% linseed oil; CAFL2.5%: obese rats fed the cafeteria diet enriched with 2.5% linseed oil; CL5%: control rats fed the standard diet enriched with 5% linseed oil; CAFL5%: obese rats fed the cafeteria diet enriched with 5% linseed oil. After verifying the normal distribution of the variables (Shapiro-Wilk test), the means of the six groups of rats were compared using a one-factor ANOVA test. This analysis was completed by the Tukey test in order to classify and compare the means in pairs. Means indicated by different letters (a, b, c) are significantly different ($p<0.05$).

II.4 Serum protein and lipoprotein levels in control and experimental rats (Figure 11, table A5 in appendices).

Serum total protein levels showed no variation between the different groups of rats studied, regardless of the type of diet.

The protein content of the different lipoprotein fractions significant variations between the different batches of rats. Protein levels in HDL were significantly higher in rats on the cafeteria diet (CAF) than in control rats (C). No other variations were noted in the other batches. Protein levels in LDL were significantly increased in rats on the cafeteria diet (CAF, CAFL2.5% and CAFL5%) compared with their respective controls (C, CL2.5%, CL5%). There was no significant difference between the control rats (C) compared to the flax control rats (CL2.5%, CL5%).

The protein content of VLDL from obese rats showed no variation compared with control rats.

II.5. Fatty acid composition of total serum lipids in different batches of rats (Table 2)

In serumsaturated fatty acid varied significantly between the groups of rats studied. Saturated fatty acid (SFA) increased in (CAF) rats compared with (C) rats, in (CAFL2.5%) rats compared with (CL2.5%) rats and in (CAFL5%) rats compared with (CL5%) rats. Enriching the cafeteria diet and the standard diet with linseed oil significantly reduced the SFA content. With a more significant and greater reduction in SFAs with the 5% linseed oil percentage.

With regard to monounsaturated fatty acids (MUFA), a decrease was observed in (CAF) rats compared with (C) rats and in (CAFL2.5%) rats compared with (CL2.5%) rats. Consumption of the flaxseed diet *resulted in* an increase serum MUFA in (CAFL2.5% and CAFL5%) compared with (CAF) rats. The increase was more pronounced with the 5% than with the 2.5% flaxseed oil percentage. However, the C18:2n-6 content showed an increase in the (C) rats compared with the (CAF) rats. The flax diet significantly reduced the C18:2n-6 both in rats on the standard diet and in obese rats on the cafeteria diet.

There was a significant increase in C18:3n-3, C20:5n-3 and C22:6n-3 in rats on the cafeteria diet or the standard diet enriched with linseed oil compared with rats on the diet not enriched with linseed oil, the increase being more marked with the 5% linseed oil percentage. However, a decrease in serum C20:4n-6 levels was noted in (CAF) compared with (C) and in (CAFL2.5 and CAFL5%) compared with (CL2.5% and CL5%) respectively.

III. TISSUE PARAMETERS Tissue parameters

III.1 Organ weights in different batches of rats (Figure 12 and Table A6)

Organ weights in experimental rats were significantly different from those in control rats. The groups of rats on the cafeteria diet (CAF, CAFL2.5% and CAFL5%) showed a significant increase in liver weight compared with their respective controls (C, CL2.5%, CL5%). However, linseed oil *caused* a significant decrease in liver weight in rats on the cafeteria diet (CAFL2.5% and CAFL5%) compared with cafeteria rats alone (CAF). The cafeteria diet *resulted in* a significant increase in adipose tissue weight in rats (CAF, CAFL2.5% and CAFL5%) compared with their respective controls (C, CL2.5%,CL5%). Enrichment of the cafeteria diet with linseed oil *resulted in* a significant reduction in adipose tissue weight in rats (CAFL2.5% and CAFL5%) compared with rats on the cafeteria diet alone (CAF), with a 1.9% reduction in rats (CAFL5%) compared with rats (CAFL2.5%). A significant reduction was also noted in the 5% linen control rats (CL5%) compared with the rats (CL2.5%). Muscle weight was significantly increased in rats on the cafeteria diet compared with rats on the standard diet. Muscle weight in rats on the cafeteria diet enriched with flaxseed oil at 2.5% or 5% (CAFL2.5% and CAFL5%) was significantly reduced compared with rats on the cafeteria diet alone (CAF). variation was observed in rats on the standard diet with or without linseed oil.

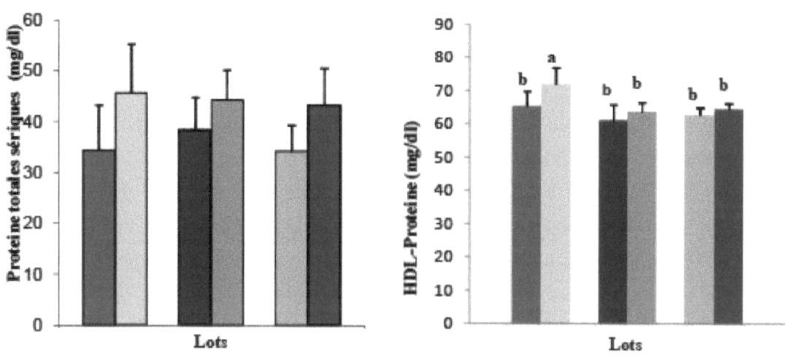

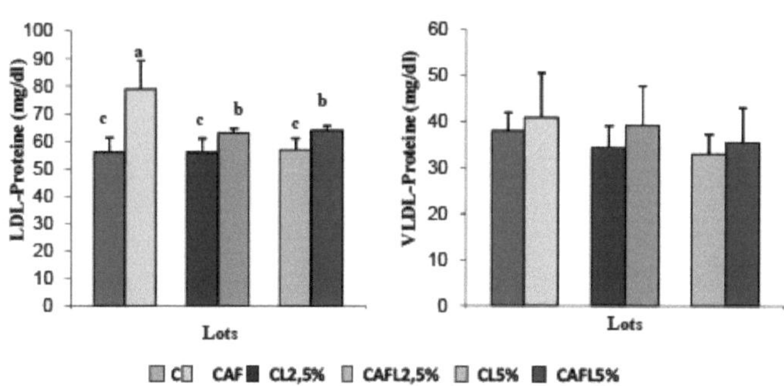

Figure 11: Protein content (mg/dl) of serum and lipoproteins in different batches of rats
Each value represents the mean ± SD, n=10.C: control rats fed the standard diet; CAF: obese rats fed the cafeteria diet; CL2.5%: control rats fed the standard diet enriched with 2.5% linseed oil; CAFL2.5%: obese rats fed the cafeteria diet enriched with 2.5% linseed oil; CL5%: control rats fed the standard diet enriched with 5% linseed oil; CAFL5%: obese rats fed the cafeteria diet enriched with 5% linseed oil. After verifying the normal distribution of the variables (Shapiro-Wilk test), the means of the six groups of rats were compared using a one-factor ANOVA test. This analysis was completed by the Tukey test in order to classify and compare the means in pairs. Means indicated by different letters (a, b, c) are significantly different ($p<0.05$).

Table 3: Fatty acid composition of serum lipids in different batches of rats

\ AG% Lots	Standard rats **(C)**	Rats cafeteria **(CAF)**	Standard flax rats2.5% **(CL2.5%)**	Rats cafeteria lin2,5% **(CAFL2,5)**	Standard rats lin5% **(CL5%)**	Rats cafeteria lin5% **(CAFL5%)**	P (ANOVA)
AGS	34.19 ± 1.3^{c}	45.97 ± 1.4^{a}	22.44 ± 1.1^{f}	38.47 ± 1.22^{b}	23.94 ± 0.33^{e}	$32,14\pm1.2^{d}$	0,0001
AGMI	19.58 ± 1.0^{a}	19.01 ± 1.0^{a}	14.23 ± 1.1^{d}	14.52 ± 1.22^{d}	15.26 ± 1.32^{c}	18.21 ± 1.6^{b}	0,0001
C18:2 n-6	18.22 ± 1.2^{a}	15.37 ± 1.2^{b}	11.62 ± 1.2^{e}	11.5 ± 1.35^{e}	14.06 ± 1.0^{c}	12.11 ± 1.0^{d}	0,0001
C18:3 n-3	$10.50\pm0.0\ 1^{d}$	9.46 ± 0.09^{e}	$28,62\pm1.2^{b}$	22.07 ± 0.24^{c}	29.66 ± 1.0^{a}	28.05 ± 0.9^{b}	0,0001
C20:4 n-6	12.01 ± 1.2^{b}	$8,3\pm1.14^{d}$	13.5 ± 2.5^{a}	$7,05\pm1.01^{d}$	$6,77\pm1.33^{c}$	$2,21\pm1.25^{e}$	0,0001
C20:5 n-3	2.17 ± 0.32^{c}	1.04 ± 0.21^{d}	$4,58\pm0.58^{b}$	4.28 ± 0.33^{b}	$5,18\pm0.56^{a}$	$4,46\pm0.63^{b}$	0,001
C22 :6	$3,31\pm0.23^{c}$	$0,85\pm0.14^{e}$	$4,51\pm0.31^{b}$	$2,11\pm0.25^{d}$	$5,13\pm0.11^{a}$	$2,81\pm0.43^{d}$	0,0001

n-3							

Each value represents the mean ± SD, n=10.C: control rats fed the standard diet; CAF: obese rats fed the cafeteria diet; CL2.5%: control rats fed the standard diet enriched with 2.5% linseed oil; CAFL2.5%: obese rats fed the cafeteria diet enriched with 2.5% linseed oil; CL5%: control rats fed the standard diet enriched with 5% linseed oil; CAFL5%: obese rats fed the cafeteria diet enriched with 5% linseed oil. After verifying the normal distribution of the variables (Shapiro-Wilk test), the means of the six groups of rats were compared using a one-factor ANOVA test. This analysis was completed by the Tukey test in order to classify and compare the means in pairs. Means indicated by different letters (a, b, c) are significantly different (p<0.05).

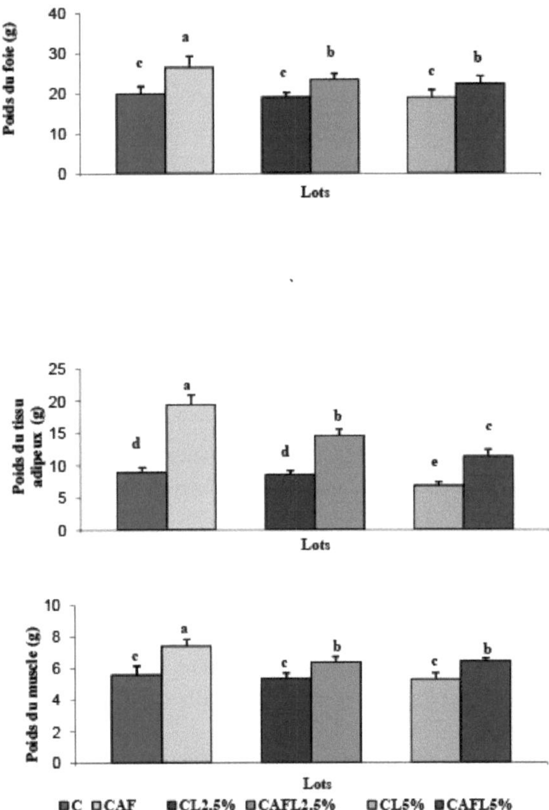

Figure 12: Organ weights in different batches of rats

Each value represents the mean ± SE, n=10.C: control rats fed the standard diet; CAF: obese rats fed the cafeteria diet; CL2.5%: control rats fed the standard diet enriched with 2.5% linseed oil; CAFL2.5%: obese rats fed the cafeteria diet enriched with 2.5% linseed oil; CL5%: control rats fed the standard diet enriched with 5% linseed oil; CAFL5%: obese rats fed the cafeteria diet enriched with 5% linseed oil. After verifying the normal distribution of the variables (Shapiro-Wilk test), the means of the six groups of rats were compared using a one-factor ANOVA test. This analysis was completed by the Tukey test in order to classify and compare the means in pairs. Means indicated by different letters (a, b, c) are significantly different ($p<0.05$).

III.2 Total lipid content (mg/g tissue) of organs in different batches of rats (Figure 13 and table A7 in appendices)

Total lipid levels in liver, muscle and adipose tissue were significantly increased in obese rats (CAF, CAFL2.5%, CAFL5%) compared with their respective controls (C, CL2.5%, CL5%). The addition linseed oil to the cafeteria diet significantly reduced the total lipid content of liver and adipose tissue in rats (CAFL2.5% and CAFL5%) compared with rats (CAF), with a more marked reduction of around 3.62% in total lipids in adipose tissue in rats (CAFL5%) compared with rats (CAFL2.5). A decrease in total lipids in the liver and adipose tissue in control rats on the standard diet enriched with linseed oil (CL2.5% and CL5%) compared with control rats on the standard diet alone (C), with a more marked decrease in total lipids in the liver in rats (CL5%) compared with rats (CL2.5). However, total lipid levels in the intestine did not differ between the obese and control batches of rats.

III. 3 Total cholesterol levels in the organs of different batches of rats (Figure 14 and table A8 in the appendices)

Liver and adipose tissue cholesterol levels were significantly increased in obese rats (CAF, CAFL2.5%, CAFL5%) compared with their respective controls (C,CL2.5%,CL5%). The addition of 2.5% and 5% linseed oil to the cafeteria diet induced a reduction in hepatic cholesterol and adipose tissue levels in rats (CAFL2.5% and CAFL5%) compared with rats (CAF).

Rats fed the cafeteria diet enriched with 5% linseed oil (CAFL5%) showed more significant reductions in adipose tissue cholesterol levels than those fed the cafeteria diet enriched with 2.5% linseed oil (CAFL2.5%). There was also a reduction in liver and adipose tissue cholesterol levels in rats (CL2.5% and CL5%) compared with control rats (C), with a greater reduction in adipose tissue cholesterol levels with the 5% linseed oil.

However, the cholesterol levels in the muscle and intestine did not vary between the different batches of rats, whatever their diet.

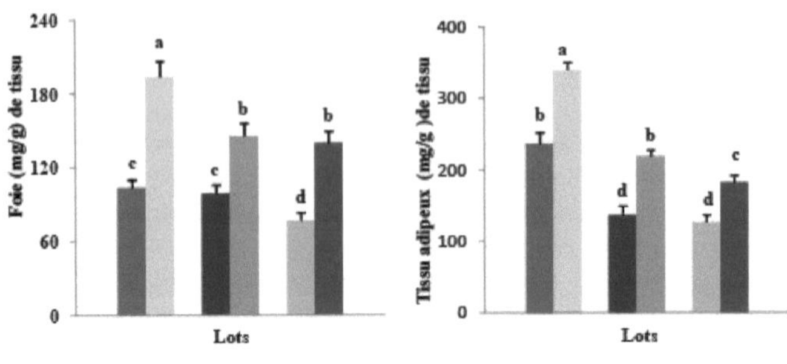

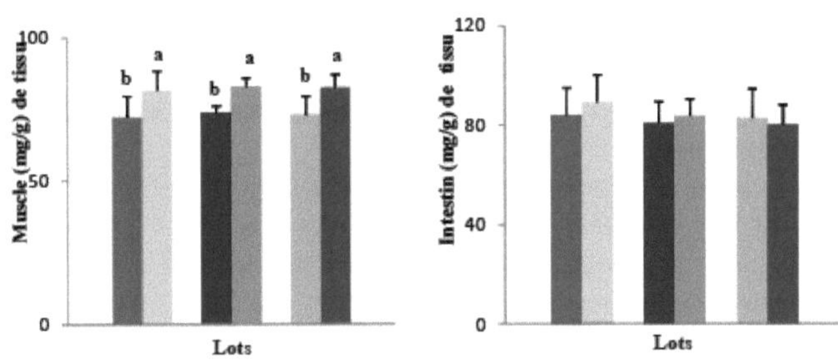

C CAF CL2,5% CAFL2,5% CL5% CAFL5%

Figure 13: Total lipid content (mg/g tissue) of organs in different batches of rats

Each value represents the mean ± SE, n=10.C: control rats fed the standard diet; CAF: obese rats fed the cafeteria diet; CL2.5%: control rats fed the standard diet enriched with 2.5% linseed oil; CAFL2.5%: obese rats fed the cafeteria diet enriched with 2.5% linseed oil; CL5%: control rats fed the standard diet enriched with 5% linseed oil; CAFL5%: obese rats fed the cafeteria diet enriched with 5% linseed oil. After verifying the normal distribution of the variables (Shapiro-Wilk test), the means of the six groups of rats were compared using a one-factor ANOVA test. This analysis was completed by the Tukey test in order to classify and compare the means in pairs. Means indicated by different letters (a, b, c) are significantly different ($p<0.05$).

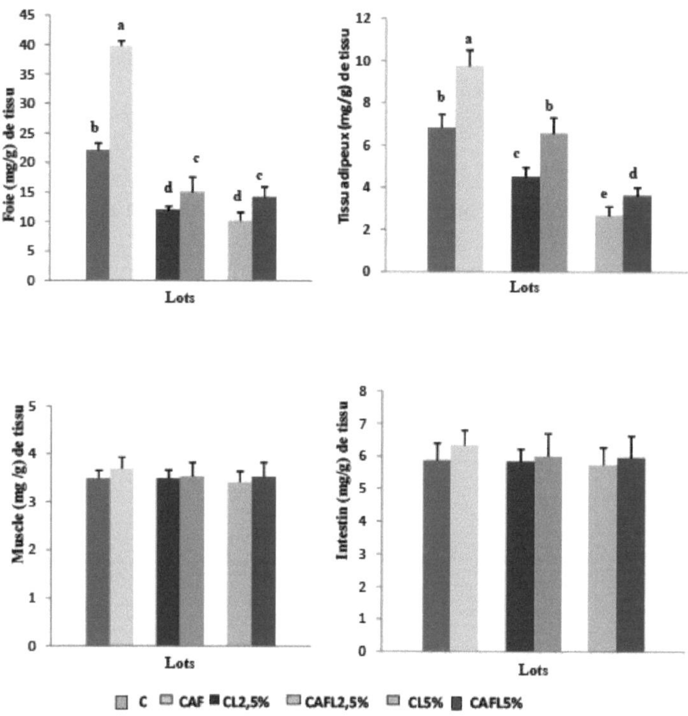

Figure 14: Total organ cholesterol levels (mg/g tissue) in different batches of rats
Each value represents the mean ± SE, n=10.C: control rats fed the standard diet; CAF: obese rats fed the cafeteria diet; CL2.5%: control rats fed the standard diet enriched with 2.5% linseed oil; CAFL2.5%: obese rats fed the cafeteria diet enriched with 2.5% linseed oil; CL5%: control rats fed the standard diet enriched with 5% linseed oil; CAFL5%: obese rats fed the cafeteria diet enriched with 5% linseed oil. After verifying the normal distribution of the variables (Shapiro-Wilk test), the means of the six groups of rats were compared using a one-factor ANOVA test. This analysis was completed by the Tukey test in order to classify and compare the means in pairs. Means indicated by different letters (a, b, c) are significantly different ($p<0.05$).

III.4 Triglyceride content of organs in different batches of rats (Figure 15, table A9 in appendices)

Obese rats on a cafeteria diet enriched or not with linseed oil (CAF, CAFL2.5%, CAFL5%) showed a significant increase in hepatic and adipocyte triglyceride levels compared with their respective controls (C, CL2.5%, CL5%).

Muscle and adipocyte hepatic triglyceride levels were significantly reduced in rats on the flaxseed oil-enriched cafeteria diet (CAFL2.5 and CAFL5%) compared with rats on the cafeteria diet alone (CAF), and this reduction was similar regardless of the percentage flaxseed oil added to the diet.

There was also a significant reduction in liver, muscle and adipose tissue triglycerides in rats on the standard diet enriched with linseed oil (CL2.5% and CL5%) compared with control rats (C).

However, no variation was observed in intestinal triglycerides in aged rats, whatever the diet.

III.5. Total protein contents (mg/g tissue) of organs in control and experimental rats (Figures 16, Table A10 in appendices).

Total protein levels in the liver and adipose tissue of aged rats varied significantly between the different batches.

Hepatic and adipocyte protein levels increased significantly in obese rats on a cafeteria diet enriched or not with linseed oil (CAF, CAL2.5% and CAFL5%) compared with their respective controls (C, CL2.5% and CL5%).

The addition linseed oil to the standard diet and to the cafeteria diet significantly reduced liver and adipocyte protein levels, with the reduction in liver protein levels being greater with the percentage of linseed oil at 5%.

However, total muscle and intestinal protein content did not vary in aged rats, whatever the diet.

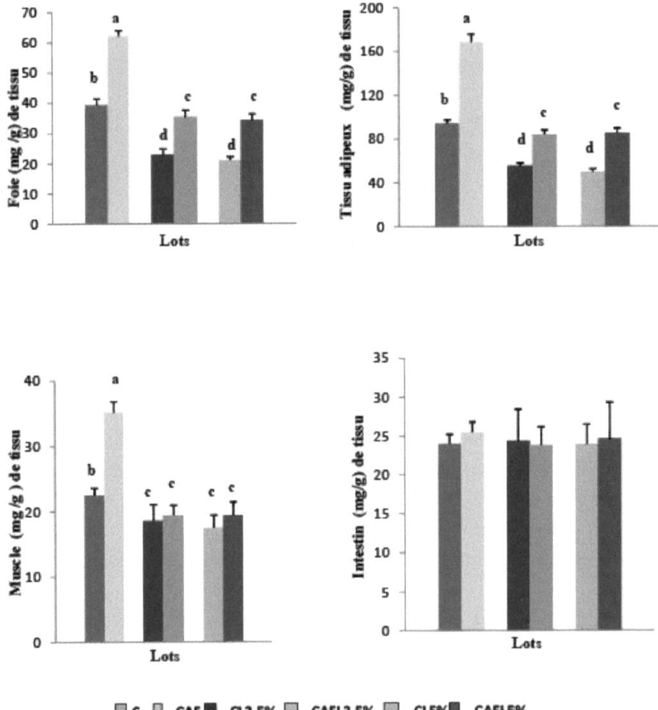

Figure 15: Total organ triglyceride levels (mg/g tissue) in different batches of rats

Each value represents the mean ± SE, n=10.C: control rats fed the standard diet; CAF: obese rats fed the cafeteria diet; CL2.5%: control rats fed the standard diet enriched with 2.5% linseed oil; CAFL2.5%: obese rats fed the cafeteria diet enriched with 2.5% linseed oil; CL5%: control rats fed the standard diet enriched with 5% linseed oil; CAFL5%: obese rats fed the cafeteria diet enriched with 5% linseed oil. After verifying the normal distribution of the variables (Shapiro-Wilk test), the means of the six groups of rats were compared using a one-factor ANOVA test. This analysis was completed by the Tukey test in order to classify and compare the means in pairs. Means indicated by different letters (a, b, c) are significantly different (p<0.05).

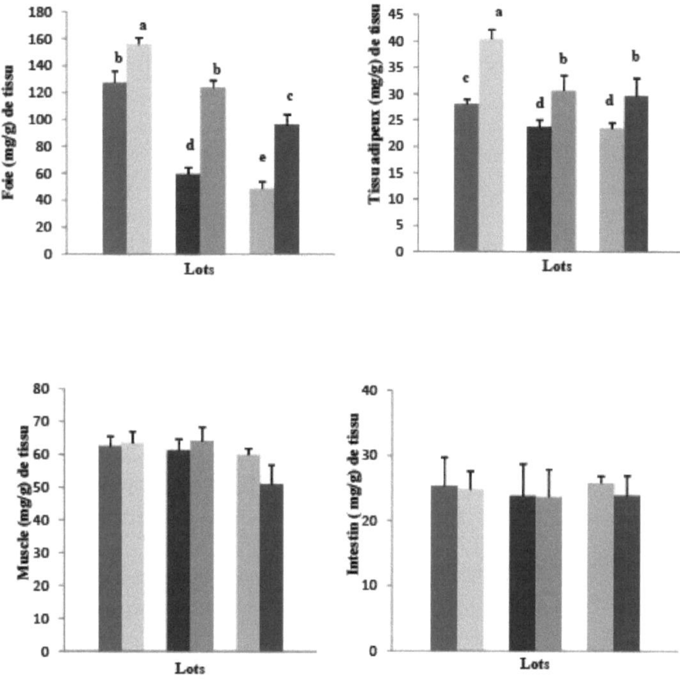

Figure 16: Total organ protein content (mg/g tissue) in different batches of rats

Each value represents the mean ± SD, n=10.C: control rats fed the standard diet; CAF: obese rats fed the cafeteria diet; CL2.5%: control rats fed the standard diet enriched with 2.5% linseed oil; CAFL2.5%: obese rats fed the cafeteria diet enriched with 2.5% linseed oil; CL5%: control rats fed the standard diet enriched with 5% linseed oil; CAFL5%: obese rats fed the cafeteria diet enriched with 5% linseed oil. After verifying the normal distribution of the variables (Shapiro-Wilk test), the means of the six groups of rats were compared using a one-factor ANOVA test. This analysis was completed by the Tukey test in order to classify and compare the means in pairs. Means indicated by different letters (a, b, c) are significantly different (p<0.05).

III.6. Tissue fatty acid composition

III.6.1. Fatty acid composition of liver lipids in different batches of rats (Table 4)

At liver level, the SFA composition of lipids showed an increase in the (CAF) rats compared with the (C) rats, in the (CAFL2.5%) rats compared with the (CL2.5%) rats and in the (CAFL5%) rats compared with the (CL5%) rats. Supplementation of the diet with linseed oil *resulted in* a significant reduction in the SFA content of liver lipids in both rats on the standard diet and obese rats on the cafeteria diet. The effect was greater with the 5% linseed oil.

The levels of hepatic MUFA and C18:2 n-6 in obese rats (CAF) are increased compared with controls (C), and in (CAFL2.5%) compared with (CL2.5%) and in (CAFL5%) compared with (CL5%).

The flax diet significantly reduced C18:2n-6 levels in obese rats (CAFL2.5% and CAFL5%) compared to (CAF) rats and also in control rats (CL2.5% and CL5%) compared to (C) rats.

The hepatic C18:3-n content was significantly reduced in rats on a diet not enriched with linseed oil compared with the other batches. Enriching the diet with linseed oil significantly increased the hepatic C18:3n-3 content in obese rats (CAFL2.5% and CAFL5%) compared with obese rats (CAF) and in rats (CL2.5% and CL5%) compared with control rats (C). A decrease in hepatic C20:4n-6 content was noted in the (CAF) rats compared with the (C) rats and in the (CAFL2.5%) rats compared with the (CL2.5%) rats and in the (CAFL2.5%) rats compared with the (CL2.5%) rats. Supplementation of the diet with linseed oil *resulted in* a significant increase in C20:4n-6 content both in the rats on the standard diet and in the obese rats on the cafeteria diet.

C20:5n-3 and C22:6n-3 levels decreased in obese rats (CAF) compared with control rats (C). Supplementing the diet with linseed oil *resulted in* a significant increase in C20:5n-3 and C22:6n-3 levels in both rats on the standard diet and obese rats on the cafeteria diet.

III.6.2. Fatty acid composition of adipose tissue lipids in different batches of rats (Table 5)

SFAs were significantly increased in obese rats (CAF, CAFL2.5% and CAFL5%) compared with their respective controls (C,CL2.5% ,CL5%).

The levels of MUFA and C18:2 n-6 in the adipose tissue of obese rats (CAF) were increased compared with controls (C).

Supplementing the diet with flaxseed oil significantly reduced the levels of MUFA and C18:2 n-6 in both rats on the standard diet and obese rats on the cafeteria diet.

The C18 :3-n content of adipose tissue was significantly reduced in rats on a diet not enriched with linseed oil compared with the other batches. Enriching the diet with linseed oil significantly increased the C18 :3n-3 content in obese rats (CAFL2.5% and CAFL5%) compared with obese rats (CAF) and in rats (CL2.5% and CL5%) compared with control rats (C).

A decrease in C20:4n-6 content was noted in (CAF) compared (C).

Supplementing the diet with flaxseed oil *resulted in* a significant increase in C20:4n-6 content in both rats on the standard diet and obese rats on the cafeteria diet.

C20:5n-3 and C22:6n-3 levels decreased in obese rats (CAF) compared with control rats (C). Supplementation of the diet with linseed oil *resulted in* a significant increase in C20:5n-3 and C22:6n-3 levels in both rats on the standard diet and obese rats on the cafeteria diet.

Table 3: Fatty acid composition of liver lipids in different batches of rats

\Lots	Standard rats	Rats cafeteria	Standard flax	Rats cafeteria	Standard flax	Rats cafeteria	P ANOVA

AG%	**(C)**	**(CAF)**	rats 2.5% **(CL2.5%)**	lin 2,5% **(CAFL2,5)**	rats 5% **(CL5%)**	linen 5% **(CAFL5%)**	
AGS	42.22 ± 1.34^{b}	47.97 ± 1.48^{a}	25.47 ± 1.41^{e}	37.67 ± 1.22^{c}	22.94 ± 1.33^{f}	31.06 ± 1.58^{d}	0,0001
AGMI	17.28 ± 1.01^{b}	18.02 ± 1.02^{a}	13.23 ± 1.05^{d}	15.52 ± 1.48^{c}	15.26 ± 1.12^{c}	19.31 ± 1.66^{a}	0,001
C18: 2n-6	19.22 ± 1.21^{b}	22.39 ± 1.11^{a}	11.62 ± 1.20^{d}	10.67 ± 1.35^{e}	13.76 ± 1.04^{c}	11.12 ± 1.08^{d}	0,0001
C18: 3n-3	0.50 ± 0.06^{d}	0.66 ± 0.07^{d}	$26{,}92\pm1.20^{b}$	23.67 ± 0.14^{c}	30.96 ± 1.04^{a}	29.73 ± 0.93^{a}	0,0001
C20:4n-6	17.8 ± 1.23^{b}	$8{,}34\pm1.04^{d}$	$15{,}58\pm2.55^{a}$	$6{,}6\pm1.11^{e}$	$8{,}69\pm1.33^{c}$	$3{,}22\pm1.25^{f}$	0,0001
C20: 5n-3	3.46 ± 0.54^{c}	1.63 ± 0.39^{d}	$4{,}25\pm0.58^{b}$	4.21 ± 0.42^{b}	$5{,}16\pm0.56^{a}$	$3{,}76\pm0.77^{c}$	0,001
C22: 6n-3	$1{,}51\pm0.24^{d}$	$0{,}91\pm0.51^{e}$	$2{,}93\pm0.12^{b}$	$1{,}66\pm0.23^{d}$	$3{,}23\pm0.14^{a}$	$1{,}8\pm0.33^{c}$	0,001

Each value represents the mean ± SE, n=10.C: control rats fed the standard diet; CAF: obese rats fed the cafeteria diet; CL2.5%: control rats fed the standard diet enriched with 2.5% linseed oil; CAFL2.5%: obese rats fed the cafeteria diet enriched with 2.5% linseed oil; CL5%: control rats fed the standard diet enriched with 5% linseed oil; CAFL5%: obese rats fed the cafeteria diet enriched with 5% linseed oil. After verifying the normal distribution of the variables (Shapiro-Wilk test), the means of the six groups of rats were compared using a one-factor ANOVA test. This analysis was completed by the Tukey tcst in order to classify and compare the means in pairs. Means indicated by different letters (a, b, c) are significantly different ($p<0.05$).

Table 4: Fatty acid composition of adipose tissue lipids in different batches of rats

Ψ-Ots AG%	Standard rats **(C)**	Rats cafeteria **(CAF)**	Standard flax rats 2.5% **(CL2.5%)**	Rats cafeteria lin 2,5% **(CAFL2,5)**	Standard flax rats 5% **(CL5%)**	Rats cafeteria linen 5% **(CAFL5%)**	P ANOVA
AGS	31.74 ± 1.22^{b}	35.62 ± 1.11^{a}	23.54 ± 1.15^{d}	27.13 ± 1.04^{c}	18.17 ± 1.23^{e}	22.70 ± 1.25^{d}	0,0001
AGMI	28.93 ± 1.06^{b}	32.33 ± 1.24^{a}	22.19 ± 1.44^{d}	24.97 ± 1.57^{c}	23.60 ± 1.67^{d}	24.41 ± 1.63^{c}	0,0001
C18: 2n-6	26.90 ± 1.41^{a}	27.80 ± 1.35^{a}	16.43 ± 1.67^{c}	17.22 ± 1.22^{c}	15.26 ± 1.81^{c}	22.07 ± 1.21^{b}	0,001
C18: 3n-3	0.89 ± 0.10^{e}	0.58 ± 0.07^{e}	29.30 ± 1.67^{b}	20.52 ± 0.08^{d}	32.08 ± 1.84^{a}	24.09 ± 0.54^{c}	0,0001
C20:4n-6	$9{,}1\pm1.72^{a}$	$2{,}3\pm0.55^{d}$	2.83 ± 1.31^{c}	$5{,}45\pm1.77^{b}$	$4{,}18\pm1.83^{b}$	2.01 ± 1.62^{c}	0,001
C20: 5n-3	0.60 ± 0.05^{c}	0.53 ± 0.04^{c}	3.21 ± 0.07^{a}	$2{,}78\pm0.06^{b}$	$3{,}88\pm0.42^{a}$	2.50 ± 0.31^{b}	0,001
C22: 6n-3	$1{,}80\pm0.15^{c}$	$0{,}83\pm0.03^{d}$	$2{,}50\pm0.11^{a}$	$1{,}93\pm0.11^{b}$	$2{,}83\pm0.25^{a}$	$2{,}02\pm0.32^{b}$	0,001

Each value represents the mean ± SE, n=10.C: control rats fed the standard diet; CAF: obese rats fed the cafeteria diet; CL2.5%: control rats fed the standard diet enriched with 2.5% linseed oil; CAFL2.5%: obese rats fed the cafeteria diet enriched with 2.5% linseed oil; CL5%: control rats fed the standard diet enriched with 5% linseed oil; CAFL5%: obese rats fed the cafeteria diet enriched with 5% linseed oil. After verifying the normal distribution of the variables (Shapiro-Wilk test), the means of the six groups of rats were compared using a one-factor ANOVA test. This analysis was completed by the Tukey test in order to classify and compare the means in pairs. Means indicated by different letters (a, b, c) are significantly different ($p<0.05$).

IV. Activities of tissue lipases and plasma LCAT

IV.1 Activities of lipoprotein lipase enzymes in the organs of different batches of rats (Figure 17 and Table A11)

Adipocyte and hepatic LPL activity was significantly increased in obese rats on the cafeteria diet (CAF, CAFL2.5%,CAFL5%) compared with their respective controls (C,CL5%,CL2.5%). Muscle LPL activity decreased significantly in obese rats on the cafeteria diet (CAF, CAFL2.5%, CAFL5%) compared with their respective controls (C, CL2.5%, CL5%). The addition linseed oil to the cafeteria diet significantly muscle LPL activity in rats (CAFL2.5% and CAFL2.5%) compared with rats (CAF), the effect being more pronounced with the 5% percentage.

IV.2 Lecithin Cholesterol Acyl Transferase (LCAT) activity in control and experimental rats (Figure 18, table A12 in appendices)

Lecithin Cholesterol Acyl Transferase (LCAT) activity was increased in rats on the cafeteria diet enriched or not with linseed oil (CAF, CAFL2.5%, CAFL5%) compared with their respective controls. LCAT activity was significantly decreased in rats on the cafeteria diet enriched with linseed oil (CAFL2.5%, CAFL5%) compared with rats on the cafeteria diet alone (CAF) and in rats on the standard diet enriched with linseed oil (CL2.5% and CL5%) compared with rats on the standard diet alone (C).

IV.3. Hormone-sensitive *lipase (HSL)* activity in control and experimental rats (Figure 19, table A13 in appendices)

Hormone-sensitive *lipase (HSL)* activity *was significantly increased in rats on the cafeteria diet alone (CAF) compared with control rats on the standard diet (C) and (CAFL2.5%) compared with rats (CAFL5%). No variation was observed in rats (CAFL5%) compared with rats (CL5%). Linseed oil significantly reduced* the activity of hormone-sensitive *lipase (HSL) in aged rats on a cafeteria diet enriched with linseed oil (CAFL2.5% and CAFL5%) compared with rats on a cafeteria diet alone (CAF), with a greater reduction with the 5% percentage (7.63%). There was also a significant decrease* hormone-sensitive *lipase (HSL)* activity *in rats on the standard diet enriched with linseed oil (CL2.5%, CL5%) compared with rats on the diet not enriched with linseed oil (C).*

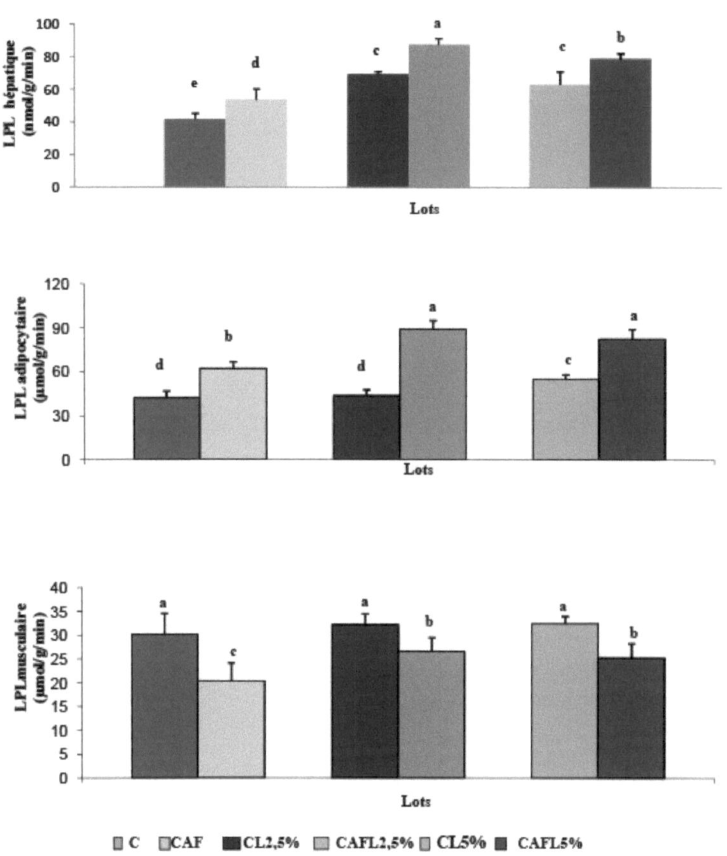

Figure 17: Lipoprotein lipase activity in different batches of rats

Each value represents the mean ± SD, n=10.C: control rats fed the standard diet; CAF: obese rats fed the cafeteria diet; CL2.5%: control rats fed the standard diet enriched with 2.5% linseed oil; CAFL2.5%: obese rats fed the cafeteria diet enriched with 2.5% linseed oil; CL5%: control rats fed the standard diet enriched with 5% linseed oil; CAFL5%: obese rats fed the cafeteria diet enriched with 5% linseed oil. After verifying the normal distribution of the variables (Shapiro-Wilk test), the means of the six groups of rats were compared using a one-factor ANOVA test. This analysis was completed by the Tukey test in order to classify and compare the means in pairs. Means indicated by different letters (a, b, c) are significantly different ($p<0.05$).

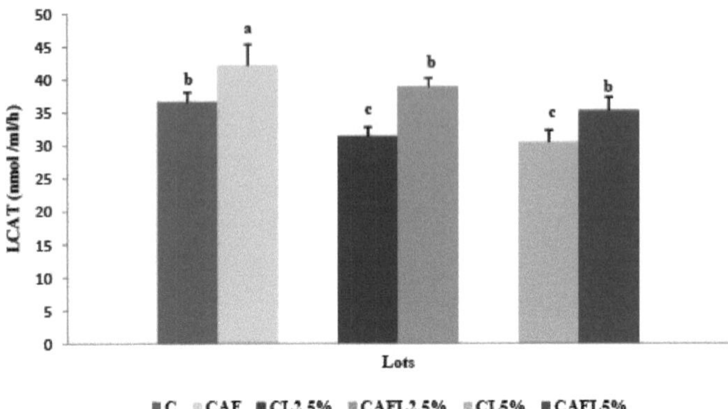

Figure 18: Lecithin cholesterol acyl transferase (LCAT) activity in different batches of rats

Each value represents the mean ± SD, n=10.C: control rats fed the standard diet; CAF: obese rats fed the cafeteria diet; CL2.5%: control rats fed the standard diet enriched with 2.5% linseed oil; CAFL2.5%: obese rats fed the cafeteria diet enriched with 2.5% linseed oil; CL5%: control rats fed the standard diet enriched with 5% linseed oil; CAFL5%: obese rats fed the cafeteria diet enriched with 5% linseed oil. After verifying the normal distribution of the variables (Shapiro-Wilk test), the means of the six groups of rats were compared using a one-factor ANOVA test. This analysis was completed the Tukey test in order to classify and compare the means in pairs. Means indicated by different letters (a, b, c) are significantly different ($p<0.05$).

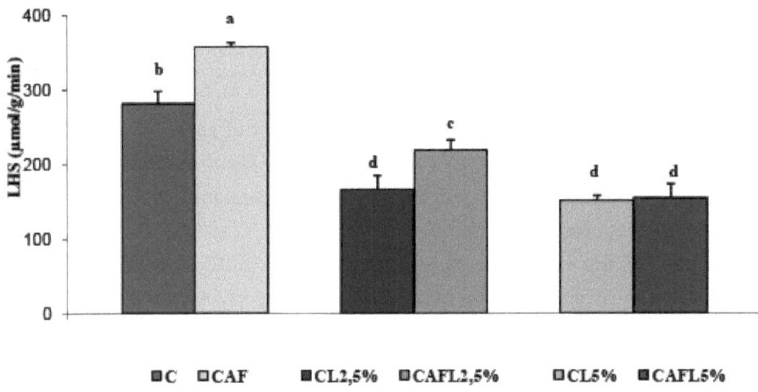

Figure 19: Hormone-sensitive lipase (HSL) activity in different batches of rats

Each value represents the mean ± SE, n=10.C: control rats fed the standard diet; CAF: obese rats fed the cafeteria diet; CL2.5%: control rats fed the standard diet enriched with 2.5% linseed oil; CAFL2.5%: obese rats fed the cafeteria diet enriched with 2.5% linseed oil; CL5%: control rats fed the standard diet enriched with 5% linseed oil; CAFL5%: obese rats fed the cafeteria diet enriched with 5% linseed oil. After verifying the normal distribution of the variables (Shapiro-Wilk test), the comparison of the mean between the six groups of rats was carried out using a one-factor ANOVA test. This analysis was completed by the Tukey test in order to classify and compare the means in pairs. Means indicated by different letters (a, b, c) are significantly different ($p<0.05$).

V . Oxidant/antioxidant status

V .1. Serum vitamin C and erythrocyte glutathione levels and activity of the erythrocyte enzyme catalase in different batches of rats (Figure 20 , Table A14 in appendices)

Serum vitamin C levels in obese rats on a cafeteria diet enriched or not with linseed oil (CAF, CAFL2.5%, CAFL5%) were significantly reduced compared with their respective controls (C, CL2.5%, CL5%).

There was a significant increase in vitamin C levels in rats (CAFL2.5% and CAFL5%) compared to rats (CAF) and in rats (CL2.5% and CL5%) compared to rats (C).

levels of reduced glutathione were significantly reduced in obese rats on a cafeteria diet enriched or not with linseed oil (CAF, CAFL2.5%, CAFL5%) compared with their respective controls (C, CL2.5%, CL5%).

Erythrocyte glutathione levels were significantly increased in (CAFL2.5% and CAFL5%) compared with (CAF) rats, and a significant increase was also noted in (CL2.5% and CL5%) compared with (C) rats. The two linseed oil percentages (2.5% and 5%) similarly increased erythrocyte GSH levels in rats on the cafeteria diet, whereas the 5% linseed oil percentage increased them more significantly in rats on the standard diet (CL5%) compared with ras (CL2.5%).

Aged rats on the cafeteria diet enriched or not with linseed oil showed a significant decrease in erythrocyte catalase activity compared with the values obtained in their respective control rats. Catalase activity increased significantly in rats fed the standard diet or the cafeteria diet enriched with linseed oil. This increase was much greater with the percentage of linseed oil at 5%.

V .2 Markers of plasma oxidative status in different batches of rats (Figure 21 and Figure 22, Table A15 in appendices)

Plasma MDA levels were significantly higher in obese subjects (CAF, CAFL2.5%, CAFL5%) compared with their respective controls (C, CL2.5%, CL5%).

Enrichment of the diet (standard or cafeteria) with linseed oil *resulted in* a significant reduction in MDA, with a reduction (7%) in rats receiving 5% linseed oil.

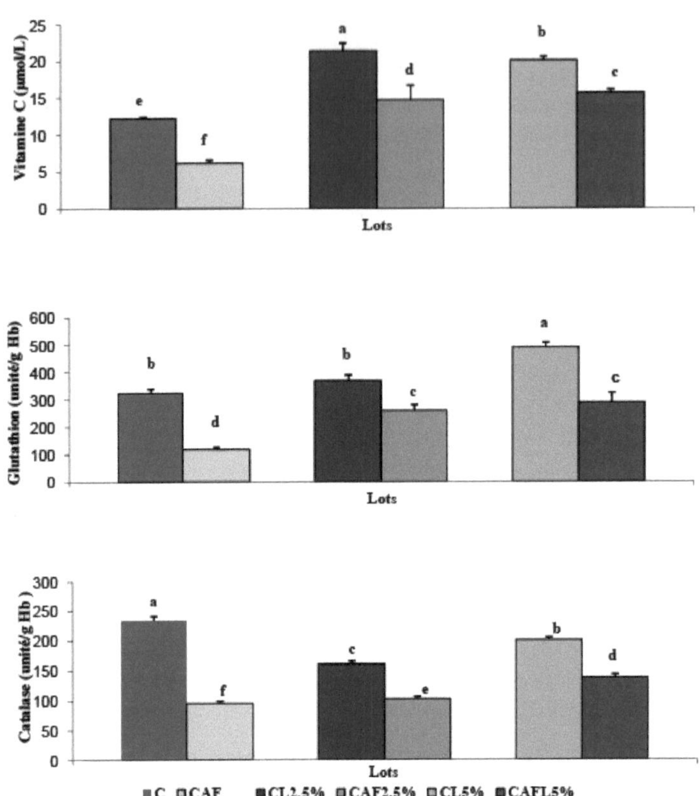

Figure 20: Serum vitamin C and erythrocyte glutathione levels and activity of the erythrocyte enzyme catalase in different batches of rats.

Each value represents the mean ± SD, n=10.C: control rats fed the standard diet; CAF: obese rats fed the cafeteria diet; CL2.5%: control rats fed the standard diet enriched with 2.5% linseed oil; CAFL2.5%: obese rats fed the cafeteria diet enriched with 2.5% linseed oil; CL5%: control rats fed the standard diet enriched with 5% linseed oil; CAFL5%: obese rats fed the cafeteria diet enriched with 5% linseed oil. After verifying the normal distribution of the variables (Shapiro-Wilk test), the means of the six groups of rats were compared using a one-factor ANOVA test. This analysis was completed the Tukey test in order to classify and compare the means in pairs. Means indicated by different letters (a, b, c) are significantly different ($p<0.05$).

Obese rats on the cafeteria diet (CAF and CAFL2.5%) showed significantly increased plasma levels of carbonylated proteins compared with their respective controls (C,CL2.5%). Enrichment of the diet with linseed oil resulted in a significant reduction in carbonylated protein levels in rats (CAFL2.5% and CAFL5%) compared with cafeteria rats alone (CAF) and in rats (CL2.5%) compared with rats on the standard diet alone (C).

Aged rats on the cafeteria diet (CAF and CAFL5%) showed a significant increase in plasma hydroperoxide levels compared with their respective controls (C, CL5%). variation was noted in rats (CAFL2.5%) compared with rats (CL2.5%).

Linseed oil significantly reduced hydroperoxide levels in rats (CAFL2.5% and CAFL5%) compared with rats (CAF) and in rats (CL2.5% and CL5%) compared with rats (C). The effect was more pronounced with the percentage of linseed oil at 5%, at 5.4%.

Conjugated diene levels (CDI) are significantly higher in (CAF, CAFL2.5% and CAFL5%) compared to (C,CL2.5%,CL5%) respectively, but significantly lower in obese (CAFL2.5% and CAFL5%) compared to (CAF) and control (CL2.5% and CL5%) compared to (C).

The rate lipoprotein oxidation was significantly increased in obese rats (CAF, CAFL2.5% ,CAFL5%) compared with their respective controls (C,CL2.5%,CL5%). Flaxseed oil significantly reduces the rate lipoprotein oxidation in obese rats.

Linseed oil did not influence the rate oxidation in the (CL2.5%) controls compared with the (C) controls, but it did reduce the rate of oxidation in the (CL5%) controls compared with the (C) controls.

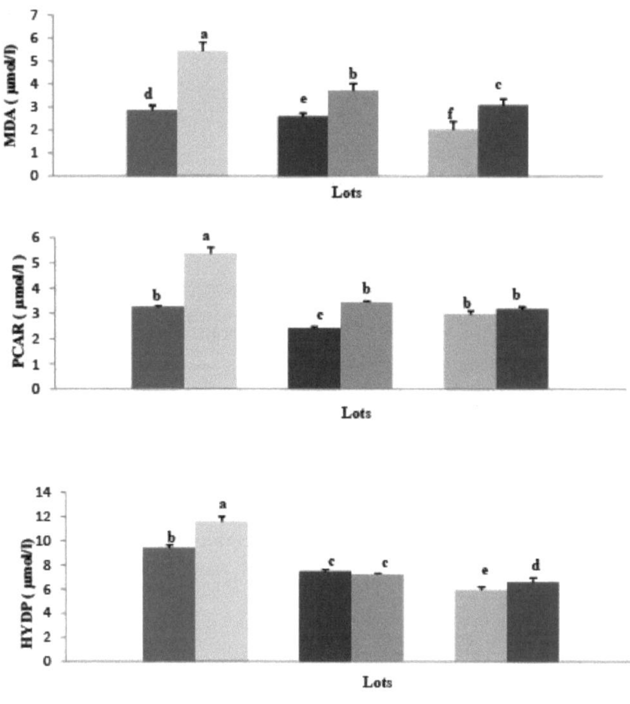

Figure 21: Markers of plasma oxidative status in different batches of rats

Each value represents the mean ± SE, n=10.C: control rats fed the standard diet; CAF: obese rats fed the cafeteria diet; CL2.5%: control rats fed the standard diet enriched with 2.5% linseed oil; CAFL2.5%: obese rats fed the cafeteria diet enriched with 2.5% linseed oil; CL5%: control rats fed the standard diet enriched with 5% linseed oil; CAFL5%: obese rats fed the cafeteria diet enriched with 5% linseed oil. After verifying the normal distribution of the variables (Shapiro-Wilk test), the means of the six groups of rats were compared using a one-factor ANOVA test. This analysis was completed by the Tukey test in order to classify and compare the means in pairs. Means indicated by different letters (a, b, c) are significantly different ($p<0.05$).

MDA: malondialdehyde; **PCAR**: carbonylated proteins, **HYDP**: hydroperoxides

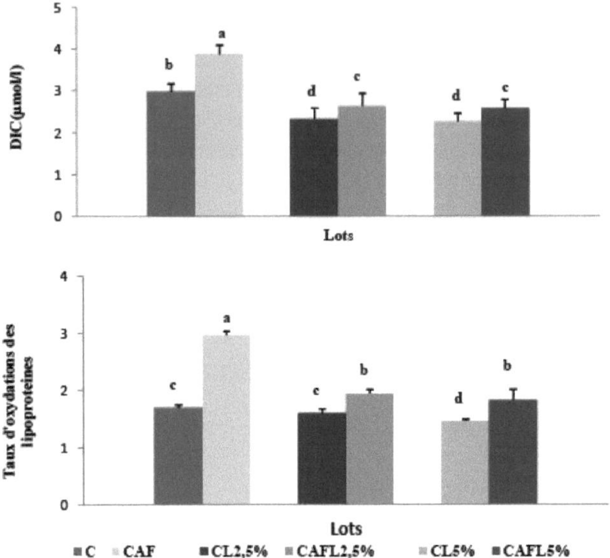

Figure 22: Markers of plasma oxidative status in different batches of rats

Each value represents the mean ± SD, n=10.C: control rats fed the standard diet; CAF: obese rats fed the cafeteria diet; CL2.5%: control rats fed the standard diet enriched with 2.5% linseed oil; CAFL2.5%: obese rats fed the cafeteria diet enriched with 2.5% linseed oil; CL5%: control rats fed the standard diet enriched with 5% linseed oil; CAFL5%: obese rats fed the cafeteria diet enriched with 5% linseed oil. After verifying the normal distribution of the variables (Shapiro-Wilk test), the means of the six groups of rats were compared using a one-factor ANOVA test. This analysis was completed by the Tukey test in order to classify and compare the means in pairs. Means indicated by different letters (a, b, c) are significantly different ($p<0.05$).

DIC: conjugated dienes

V.3 Markers of the oxidative/antioxidant status of the liver in the different batches of rats (Figure 23 and Table A16 in the appendices)

Hepatic malondialdehyde levels in obese rats (CAF, CAFL2.5% and CAFL5%) are significantly increased compared to their respective controls (C, CL2.5%, CL5%).

On the other hand, linseed oil supplementation of the cafeteria diet and the standard diet *resulted in* a reduction in hepatic malondialdehyde levels in (CAFL2.5% and CAFL5%) rats compared with (CAF) rats and in (CL2.5% and CL5%) rats compared with (C) rats. The reduction was similar whatever the percentage of linseed oil.

Analysis of hepatic carbonyl proteins showed a significant increase in rats on the cafeteria diet enriched or not with linseed oil compared with their respective controls. Supplementation of the diet with linseed oil *led to* a reduction in the levels of in hepatic carbonyl proteins in both control rats consuming the standard diet and obese rats on the cafeteria diet. The effect was more pronounced at 5% linseed oil than at 2.5%, at 5.3%. Erythrocyte hepatic glutathione levels were significantly increased in obese rats (CAF, CAFL2.5%, CAFL5%) compared with their respective controls (C, CL2.5%, CL5%). Supplementation of the diet with linseed oil *resulted in* a significant decrease in hepatic glutathione levels in rats consuming the cafeteria diet, and this decrease was similar regardless of the linseed oil percentage.

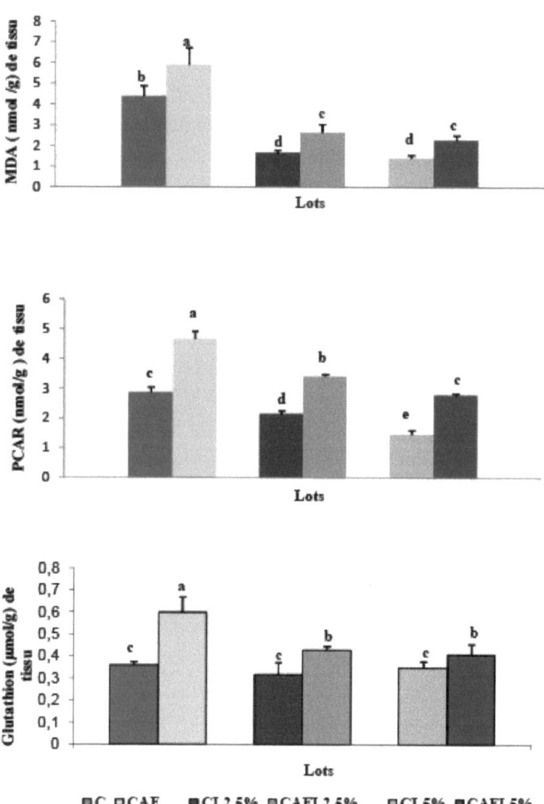

Figure 23 Markers of liver oxidative/antioxidant status in different batches of rats Each value represents the mean ± ES, n=10.C: control rats fed the standard diet; CAF: obese rats fed the cafeteria diet; CL2.5%: control rats fed the standard diet enriched with 2.5% linseed oil; CAFL2.5%: obese rats fed the cafeteria diet enriched with 2.5% linseed oil; CL5%: control rats fed the standard diet enriched with 5% linseed oil; CAFL5%: obese rats fed the cafeteria diet enriched with 5% linseed oil. After verifying the normal distribution of the variables (Shapiro-Wilk test), the means of the six groups of rats were compared using a one-factor ANOVA test. This analysis was completed by the Tukey test in order to classify and compare the means in pairs. Means indicated by different letters (a, b, c) are significantly different ($p<0.05$).

MDA: malondialdehyde; **PCAR:** carbonylated proteins

V .4. Markers of oxidant/antioxidant status of adipose tissue in different batches of rats (Figure 24 and Table A17 in appendices)

MDA levels in the adipose tissue of aged rats on the cafeteria diet supplemented or not with linseed oil were significantly elevated compared with their respective controls. MDA levels in adipose tissue decreased significantly in rats on the cafeteria diet and on the standard diet supplemented with linseed oil, with a similar decrease whatever the linseed oil percentage.

The levels of carbonylated proteins in adipose tissue were significantly increased in rats (CAF, CAFL2.5%) compared with their respective controls (C, CL2.5%). No variation was observed between rats (CAFL5%) and rats (CL5%). Linseed oil significantly reduced carbonylated protein levels in rats on the cafeteria and standard diets.

The values obtained for glutathione levels in adipose tissue showed a significant increase in rats on the cafeteria diet with or without flaxseed oil compared with their respective controls. Supplementation of the cafeteria diet with linseed oil significantly reduced glutathione levels in adipose tissue. There was no significant difference between rats on the standard diet enriched with linseed oil and rats on the standard diet not enriched with linseed oil.

V .5. Markers of muscle oxidant/antioxidant status in different batches of rats (Figure 25 and table A18 in appendices)

Muscle MDA levels increased significantly in rats on the cafeteria diet enriched or not with linseed oil compared with their respective controls. Linseed oil significantly reduced muscle MDA levels in rats on the cafeteria and standard diets, with the effect of linseed oil being similar whatever the percentage.

Muscle PCAR levels were significantly higher in the obese batches compared with their control counterparts. Consumption of the flaxseed oil-enriched diet *led to* a reduction in PCAR levels in both the rats on the standard diet and the obese rats on the cafeteria diet.

Muscle glutathione levels were unchanged in experimental rats compared with control rats.

V .6. Markers of gut oxidant/antioxidant status in different batches of rats (Figure 26 and Table A19)

There was no difference in MDA, GSH or PCAR levels between the different groups of rats, whatever their diet.

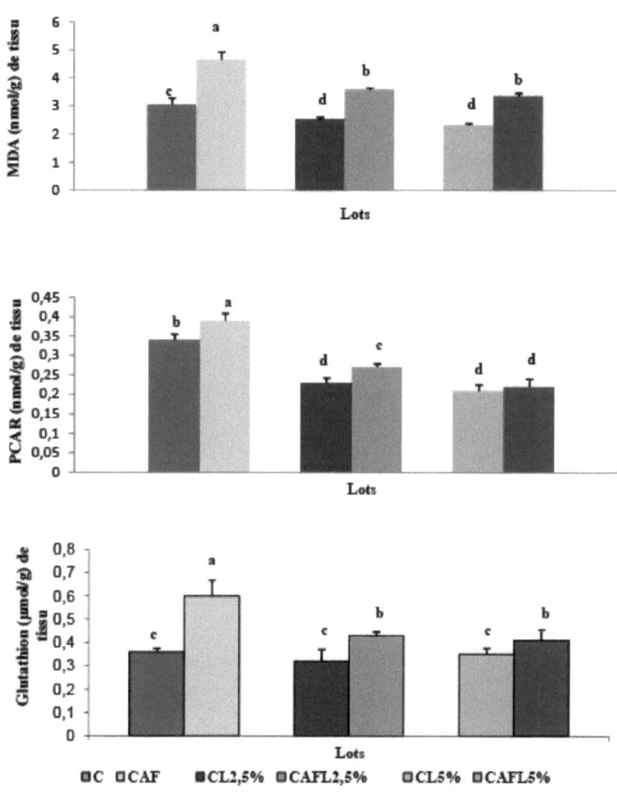

Figure 24: Markers of oxidant/antioxidant status of adipose tissue in different batches of rats.

Each value represents the mean ± SD, n=10.C: control rats fed the standard diet; CAF: obese rats fed the cafeteria diet; CL2.5%: control rats fed the standard diet enriched with 2.5% linseed oil; CAFL2.5%: obese rats fed the cafeteria diet enriched with 2.5% linseed oil; CL5%: control rats fed the standard diet enriched with 5% linseed oil; CAFL5%: obese rats fed the cafeteria diet enriched with 5% linseed oil. After verifying the normal distribution of the variables (Shapiro-Wilk test), the means of the six groups of rats were compared using a one-factor ANOVA test. This analysis was completed by the Tukey test in order to classify and compare the means in pairs. Means indicated by different letters (a, b, c) are significantly different ($p<0.05$).

MDA: malondialdehyde; **PCAR:** carbonylated proteins

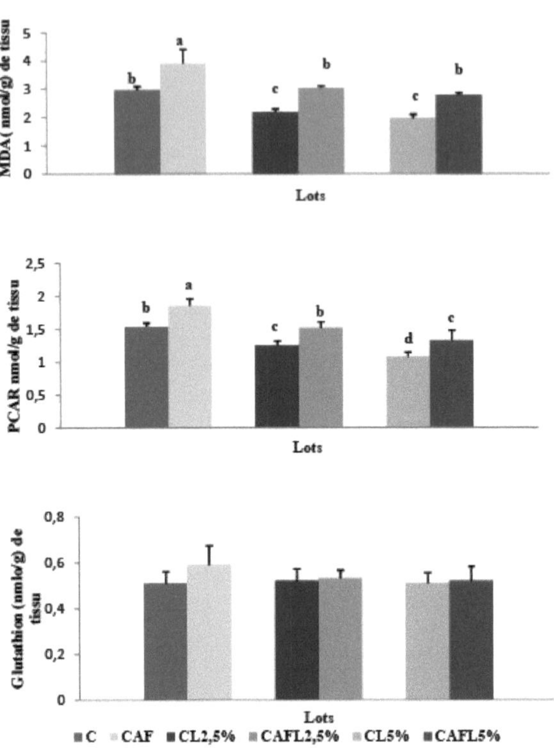

Figure 25: Markers of muscle oxidant/antioxidant status in different batches of rats

Each value represents the mean ± SD, n=10.C: control rats fed the standard diet; CAF: obese rats fed the cafeteria diet; CL2.5%: control rats fed the standard diet enriched with 2.5% linseed oil; CAFL2.5%: obese rats fed the cafeteria diet enriched with 2.5% linseed oil; CL5%: control rats fed the standard diet enriched with 5% linseed oil; CAFL5%: obese rats fed the cafeteria diet enriched with 5% linseed oil. After verifying the normal distribution of the variables (Shapiro-Wilk test), the means of the six groups of rats were compared using a one-factor ANOVA test. This analysis was completed by the Tukey test in order to classify and compare the means in pairs. Means indicated by different letters (a, b, c) are significantly different ($p<0.05$).

MDA: malondialdehyde; **PCAR:** carbonylated proteins

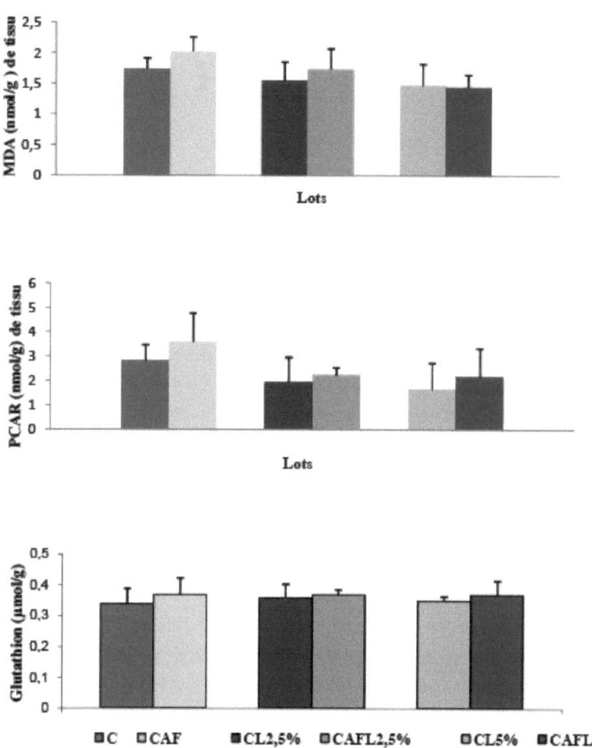

Figure 26: Markers of oxidant/antioxidant status in the intestine of different batches of rats.

Each value represents the mean ± SD, n=10.C: control rats fed the standard diet; CAF: obese rats fed the cafeteria diet; CL2.5%: control rats fed the standard diet enriched with 2.5% linseed oil; CAFL2.5%: obese rats fed the cafeteria diet enriched with 2.5% linseed oil; CL5%: control rats fed the standard diet enriched with 5% linseed oil; CAFL5%: obese rats fed the cafeteria diet enriched with 5% linseed oil. After verifying the normal distribution of the variables (Shapiro-Wilk test), the means of the six groups of rats were compared using a one-factor ANOVA test. This analysis was completed by the Tukey test in order to classify and compare the means in pairs. Means indicated by different letters (a, b, c) are significantly different ($p<0.05$).
MDA: malondialdehyde; **PCAR:** carbonylated proteins

CHAPTER 4

DISCUSSION

Obesity, a major contributor to morbidity and mortality, is the most common metabolic disease in the world, so much so that its prevention has become one of the WHO's priorities (LE GOFF et al., 2008; DIDIER et al., 2009). A modern-day pandemic, it part of the problem of chronic and systemic diseases (SCHLIENGER et al., 2010), and predisposes individuals to an increased risk of developing numerous diseases, including type II diabetes, arterial hypertension, coronary heart disease, dyslipidemia and cancer (VERGES, 2001; VINER et al., 2005; AUBIN, 2009). It has also been observed that these metabolic diseases induce disorders of the antioxidant system; the reactive oxygen species (ROS) produced during basal metabolism are neutralised by various defence systems. However, in the course of obesity, their increased production and/or a reduction in the anti-radical defence systems leads to oxidative stress causing tissue damage caused by the attack of ROS on target molecules such as lipids, proteins and DNA (HIGDON FREI, 2003; CODONER-FRANCH et al., 2004; MERZOUK et al., 2004). Ageing is also associated with lipid disorders and an increase in oxidative stress (KREGEL et al., 2007). Given that the elderly are already at risk of oxidative stress, the additional metabolic disturbances caused by obesity increase their susceptibility to the latter. In order to combat obesity and the consequences of ageing, and to reduce and slow their progression, it is vital to try and understand the mechanisms behind their development. Nutritional prevention is therefore one of the strategies used to prevent the development of obesity and ageing, through special diets. Several studies have focused on n-3 polyunsaturated fatty acids and their health benefits (DELORGERIL et al., 1994 ; NORDOY et al., 2001). Among the plants richest in n-3 PUFAs is linseed Linum usitatissimum (Linaceae), (CARTER et al., 1993) which has been consumed for centuries for its good flavour and its range of nutritional benefits revealed by scientific research (OOMAH et al., 2001). In addition, linseed oil contains 53.3% α-linolenic acid (C18:3 n-3) and 12.7% linoleic acid (C18:2n-6), giving it the highest n-3/n-6 ratio of all plant sources (TZANG et al., 2009). After elongation and desaturation, α-linolenic acid (ALA) yields eicosapentaenoic acid (EPA) and docosahexaenoic acid (DHA), which may have beneficial effects on health and the control of chronic diseases (VIJAIMOHAN et al., 2006). Clinical conditions such as cardiovascular disease, arterial hypertension, cancer, skin disease and immune disorders such as renal failure, rheumatoid arthritis, arthritis and osteoporosis are among the most common chronic diseases in which DHA may have beneficial effects.

and multiple sclerosis can be prevented by the ALA contained in linseed oil (KELLEY et al., 1991; POUDYAL et al., 2013).

The intake of n-3 PUFAs has also been associated with positive effects on obesity and the modulation of age-related oxidative stress. In fact, a diet rich in polyunsaturated fatty acids could reduce the production of free radicals and considerably improve health during ageing by regulating antioxidant enzyme genes and improving the body's antioxidant defence systems (HARRIS, 1989; KESAVULU et al., 2002; MERZOUK and KHAN, 2003).

It is difficult to monitor metabolic changes in humans because of the length of the various life stages. We therefore used an animal model of nutritional obesity, the aged Wistar rat, in which consumption of the cafeteria diet induces obesity (BOUANANE et al., 2009). Our study dealt with two main aspects:

1) Determining the effects of the cafeteria regime on :

> Alterations in lipid metabolism by studying blood parameters (total cholesterol, triglycerides and proteins in plasma and lipoprotein fractions), the protein and lipid composition of organs (liver, adipose tissue, muscle, intestine) and the activity of enzymes involved in lipid metabolism: LPL, LHS and LCAT.

> Alterations in oxidant/antioxidant status by measuring oxidant status parameters: malondialdehyde, hydroperoxides and carbonylated proteins and antioxidant status parameters: Vitamin C, as well as catalase activity and glutathione.

2) Study of the effect of 2.5% and 5% n-3 PUFA-rich linseed oil supplementation of the cafeteria diet on the major metabolic disturbances (dyslipidaemia and oxidative stress) caused by obesity in aged rats during two months of experimentation.

Models of obesity have been developed in animals, using various types of diet such as high glucose, high fat and Western (combined glucose/fat) diets (AUBIN, 2009). These models provide a method for determining the many underlying mechanisms contributing to the development of pathologies associated with obesity. Among energy-rich diets, hyperlipidic diets are more likely to lead to weight gain than high-sugar diets (BOOZER et al., 1998). In order to generate significant weight gain in rats and thus provide a good study model for obesity, we subjected the animals to a so-called "cafeteria" diet. This diet is one of the experimental models of nutritional obesity currently available. It is a hyperlipidic and hypercaloric diet, and has the advantage of being similar to the majority of human cases in which obesity is triggered by the deliberate overconsumption of foods rich in lipids and calories (DARIMONT et al., 2004). Authors have established that the main cause of obesity is an excess of calorie intake associated with a reduction in physical expenditure (DAUBRESSE et al., 2005 ; PASQUET et al., 2007). However, the increase in calorie intake does not appear to be the only important factor in the genesis of obesity; the nature of macronutrients plays a primordial role in the accumulation of fat mass; for an isocaloric intake, a meal rich in carbohydrates and very low in lipids is less adipogenic than a meal rich in lipids and normal in carbohydrates (LISSNER and HEITMANN, 1995). In fact, dietary fats have a higher energy value than other macronutrients, low satiety, high energy density and high palatability (DREWNOWSKI, 1994), which may explain why a high-fat diet can lead to an increase in energy intake, resulting in a long-term increase in body fat (BLUNDELLE and KING, 1996). This may explain the increased weight of our experimental rats consuming the hyperlipidic, high-calorie "cafeteria" diet compared with their controls consuming the standard diet. This is in agreement with the work of BENKALFAT et al (2011) on young rats. MILAGRO et al (2006) who indicated that a hyperlipidic diet in Wistar rats induced an increase in food intake and body weight with an accumulation of lipids in adipose tissue. Adipose tissue was naturally the first target of the researchers, since it stores fat. White adipose tissue plays a central role in lipid homeostasis and in maintaining energy balance. The main cellular components of adipose tissue, adipocytes, store energy in the form of triglycerides or release it in the form of fatty acids, depending on the individual's nutritional status. The development of adipose tissue results from the differentiation of precursor cells or preadipocytes present in the non-adipose fraction or stroma-vascular fraction (GESTA et al., 2007). The transition from preadipocytes to differentiated adipocytes is influenced by glucocorticoids, and triglyceride accumulation is mainly orchestrated by two categories of transcription factors belonging to the PPAR (Peroxisome proliferator activated receptor) family and the C/EBP (Echancer binding proteins) family (GREGOIRE et al., 1998; ROSEN and MAC DOUGALD, 2006). Adipocytes regulate energy metabolism by secreting inflammatory compounds, including cytokines such as TNFa and IL-6, and hormones such as leptin (TILG and MOSCHEN, 2006). Many authors define obesity as a condition characterised by adipocyte hypertrophy and/or hyperplasia leading to an alteration in the morphology and secretory function of adipose tissue (HIRSCH and BATCHELOR, 1976; ANDRE, 2008). Our results show a significant increase in the relative weight of adipose tissue and total lipids in rats consuming the cafeteria diet compared with controls. Visceral fat accumulation is an important marker of adipose tissue dysfunction (DESPRES and LEMIEUX, 2006; GAUVREAU et al., 2011), and is due to an inability to cope with excess food intake (BASDEVANT, 2006). This is in line with the results of BOUANANE et al (2009) who observed an increase in adipose tissue weight in young rats on the cafeteria diet compared with control rats. These results are also in line with the work of PETIT et al (2007) who showed that a high-fat diet leads to increased cell proliferation in adipose tissue in rats and considerably affects the physiology and intestinal use of lipids, obesity through hyperphagia.

Our results also show that the relative weights of liver and muscle tissue show a significant increase in aged rats on the cafeteria diet compared with their controls. This may be due to the fat enrichment provided by the cafeteria diet, since the liver acts as a regulator of the production of energy substrates, adipose tissue as a fat storage site and muscle as the major site of use of these energy sources, and all the interactions between these tissues are partly controlled by insulin. The dialogue between the organs involved in controlling the energy balance is one of the pathophysiological factors in obesity (BASDEVANT, 2006). Adipose tissue has endocrine, paracrine and even immune functions that affect other tissues such as the liver and skeletal muscle (GAUVREAU et al., 2011). Consequently, its imbalance can influence the functioning other systems (BOUHALI, 2006). Our results are in line with the work of MILAGRO et al (2006) who showed that total liver weight increased by 36% in rats on the cafeteria diet compared with a control group. Our results show a significant increase in total liver lipids in rats on the cafeteria diet compared with control rats. The cafeteria diet is a good model of insulin resistance causing lipid accumulation in the liver (MILAGRO et al., 2006).

However, analysis of the lipid composition of the organs suggests that there are alterations in aged rats on the cafeteria diet compared with control rats: total cholesterol and triglyceride levels in the liver and adipose tissue are significantly higher in aged rats on the cafeteria diet compared with controls. Due to the low oxidative autoregulation capacity of lipid substrates, excess lipid intake is mainly stored in adipose tissue, leading in the long term to an increase in adipose mass and its dysfunction, with dramatic consequences in various tissues (GAUVREAU et al., 2011).

KIM et al (2000) have established that the accumulation of triglycerides in muscle and liver is linked to insulin resistance, resulting a disturbance in insulin signalling in these tissues. This may explain the significant increase in hepatic, muscle and adipose tissue triglyceride (TG) levels in aged rats on the cafeteria diet compared with controls. The increase in hepatic cholesterol synthesis in rats on the cafeteria diet compared with controls may be interpreted by the excess acetyl-CoA resulting from the hyperglycaemia observed in these rats. In the intestine, TG and total cholesterol (TC) levels did not vary between rats on the cafeteria diet and their controls. According to the UNGER team (2003), only adipose tissue is designed to accumulate TGs (enzymes and receptors specific for this storage); if other tissues accumulate them (ectopic fat), their function would be impaired. Lipotoxicity results an ectopic accumulation of lipids in the liver, but also the muscle will be involved in the insulin resistance of these different tissues (DESPRES and LEMIEUX, 2006). Insulin resistance (IR) is characterised by a reduction in the cellular and tissue response to insulin in the presence a normal concentration insulin or as a normal response to hyperinsulinism (BODEN AND SHULMAN 2002; MCGARRY, 2002). Insulin is a protein hormone secreted by the β cells of the islets of Langerhans in the pancreas. Its functions are exerted on many cells in the body, with the exception of nerve cells, and its main targets are muscle, adipocytes and the liver. It is mainly responsible for the transmembrane transport (via GLUTs) and breakdown of glucose in cells. More specifically in adipose tissue, it stimulates lipogenesis and inhibits lipolysis. Adipose tissue is a special tissue because it is both the victim and the culprit of IR. The distribution of fat (visceral or subcutaneous) has different consequences for IR, with visceral fat being the main cause of IR. Indeed, the accumulation of visceral fat increases the release GLA into the portal vein and decreases the production of adiponectin, leading to a reduction in insulin sensitivity (YOUSSEF, 2009). Conversely, subcutaneous fat is metabolically less active than visceral fat due to its lower density of β-adrenergic receptors and higher density of α-adrenergic receptors (CARLSON et al., 1969). In the liver (which is responsible for 75 to 85% of glucose production in the post-absorptive phase) (STUMVOLL et al, 1997), IR is manifested by increased oxidation of FFAs which stimulate neoglucogenesis and triglyceride synthesis, resulting in overproduction of glucose which contributes to deterioration in glucose tolerance and promotes hyperglycaemia in June (GASTALDELLI et al., 2000). This may explain the hyperglycaemia observed in aged rats on the cafeteria diet compared with controls. Metabolic disorders due to obesity may be aggravated during ageing, where age-related lipid alterations contribute to insulin resistance (HEBUTERNE et al., 2001).

FERRANNINI et al (1983) and GROOP et al (1991) have established that in peripheral tissues other than the liver, FFA compete with glucose, slowing down its uptake and oxidative use, thereby increasing circulating glucose levels. FFAs could also affect intracellular glucose transport by interfering with insulin signalling. Increased muscle concentrations of certain fatty acid metabolites such as diacylglycerol and acyl-coA are thought to stimulate phosphorylation of insulin receptors or insulin receptor substrates (IRS1) via protein kinase C, thereby inhibiting insulin signalling mechanisms, resulting in reduced glucose transport and hyperglycaemia (SHULMAN et al., 2000; YU et al., 2002). Poor fat distribution is thought to play a major role in the development of IR, delaying insulin clearance and increasing lipid synthesis. Lipid synthesis depends on lipoprotein lipase, an enzyme synthesised by many tissues, and its main function is to release free fatty acids into the bloodstream from the hydrolysis of TG in chylomicrons and VLDL (BRAUN and SEVERSON, 1992; GOLDBERG, 1996). Our results show an increase adipocyte LPL activity in aged rats on the cafeteria diet compared with their controls, which is in agreement with studies by SCHWARTZ and BRUNZELL (1981); SADUR et al. (1984); BESSESEN et al. (1991); BENKALFAT et al. (2011) who have shown that obesity in humans and rodents is characterised by an increase in LPL expression by the adipocyte, which facilitates the synthesis of TGs from GLA, and thus contributes to the excess of adipose tissue. The increase adipocyte LPL activity in experimental rats may also be due to the hyperglycaemia observed in these rats. Indeed, ONG and KERN. (1989) have shown that glucose in vitro increases the activity and synthesis of LPL by adipocytes in culture. The poor distribution of fat between adipose tissue, muscle and liver is thought to play a major role in the development of IR. When adipocytes and myocytes share the same culture medium, muscle cells develop IR due to phosphorylation of serine/threonine (Akt) kinases. The consequence is a reduction in GLUT- 4 translocation and therefore glucose uptake (DIETZE et al., 2002; DIETZE et al., 2005). This is associated with an overexpression of muscle LPL in the animal (FERREIRA et al., 2001). Our results show a reduction in muscle LPL activity in aged rats on the cafeteria diet compared with controls. This is in line with other studies showing that LPL activity in skeletal muscle is markedly reduced following weight loss (ECKEL et al., 1995; ROBERT and WANG, 2009). Hepatic LPL is significantly increased in obese rats compared with controls. This may be linked to a state of insulin resistance. KIM et al (2000) demonstrated that overexpression of LPL in the liver induced a state of glucose intolerance and reduced suppression of hepatic glucose production. The main role of LPL in peripheral tissues is to supply fatty acids for storage and/or oxidation. In obesity, the increase in lipogenesis is also accompanied by an alteration in the lipolytic function of adipose tissue. LARGE et al (1998) have shown that there is a strong correlation between lipolytic capacity and the level expression of hormone-sensitive lipase (HSL). LHS is an enzyme capable of hydrolysing TG, diglycerides and cholesterol (LANGIN et al., 2000). However, in the white adipocyte, its main activity is the hydrolysis of triglycerides and diglycerides. Its activity is regulated by phosphorylation of serine residues (MAZZUCOTELLI and LANGIN, 2006). Insulin induces dephosphorylation of LHS, leading inactivation of the enzyme (MIYOSHI et al., 2006). Our results show an increase in LHS activity in obese rats compared with controls, possibly due to adipocyte insulin resistance, where insulin has little influence on lipolysis (Jocken et al., 2007). Indeed, according to some authors, obese rats are characterised by whole-body insulin resistance with beta cell dysfunction in adulthood (MERZOUK et al., 2001). Our results contradict those of LANGIN (2005) who observed a relationship between a decrease in stimulated lipolytic capacity and a decrease in LHS expression in adipocytes from differentiated obese subjects in primary culture. A defect in LHS expression could constitute an early event in the development of obesity, aimed protecting the body against excessive release fatty acids. Obesity, ageing and insulin resistance are strongly associated with quantitative and qualitative changes in plasma lipids and lipoproteins (DENK, 2001; VERGES, 2001). During IR, FFAs increase in the circulation due to the inability of insulin to inhibit lipolysis and the reduction in peripheral elimination of FFAs (KISSEBAH and PEIRIS, 1989). This increase in FFA and glucose levels increases the concentration of circulating TGs (GOLAY et al., 1987). Our results are in line with these studies, since high plasma levels of TG, TG-VLDL and TG-LDL were observed in experimental rats compared with controls. For GOTTO. (1998), hypertriglyceridaemia is linked to hepatic overproduction of VLDL and a reduction in its metabolic clearance. Our results also show an increase in TG-HDL in experimental rats consuming the hypercaloric hyperlipidic diet compared with their controls consuming the standard diet. For some authors, the amount of lipid in the food seems to have no effect on plasma TG concentration (SCHWARZ et al., 1981).

Experimental rats consuming the high-fat diet also showed hypercholesterolaemia associated with a highly significant rise in CVLDL and LDL-C levels compared with controls. Many authors have established that an increase in the lipid content of the food *leads to* an increase in the concentration of plasma cholesterol, LDL-cholesterol in guinea pigs (FERNANDEZ et al., 1996; ROMERO and FERNANDEZ, 1996), hamsters (BENNET et al., 1995; SALTER et al., 1998) and humans (DREON et al., 1998; VIDON et al., 2001). Other studies have shown that in guinea pigs, increasing the lipid content of the feed alters the composition of plasma lipoproteins, in particular by increasing the proportion of cholesterol esters in VLDL and LDL, These changes in lipoprotein composition are associated with an increase in the activities of hepatic 3-hydroxy-3- methylglutaryl coenzyme A reductase (an enzyme involved in cholesterol synthesis), hepatic ACAT and plasma LCAT, which is in agreement with our results showing an increase in plasma LCAT in aged rats on the cafeteria diet compared with control rats (FERNANDEZ et al., 1995; MAGKOS et al., 2009). In this same species, a slower disappearance of plasma LDL was also observed, as well as a reduction in the number of hepatic LDL receptors, which could explain the hypercholesterolaemia induced by a high-fat diet (FERNANDEZ et al., 1995; FERNANDEZ et al., 1996). However, the molecular mechanisms responsible for the regulation of the lipoprotein profile by the lipid content of the food remain to be clarified. According to LOUGHEED and STEINBRECHER (1996), high plasma levels of LDL-C are due to excessive hepatic secretion of lipoproteins or to a defect in the elimination of LDL, which favours the entrapment of LDL in the intima and its oxidation by free radicals generated by adjacent cells. In addition, experimental rats on the cafeteria diet have low levels of HDL-C compared with controls. This reduction is probably due to the increased synthesis of TG-VLDL, which drains cholesterol esters and Apo AI from HDL (VERGES, 2001), which may also explain the increase in thc protein content of HDL in experimental rats (CHAPMAN, 1982; SCHAEFER and LEVY, 1985).

The fatty acid composition of serum is altered, with an increase in the weight percentages of SFAs in serum, liver and adipose tissue. This increase in SFAs may be due to excess intake on the cafeteria diet or to an increase in their synthesis. A high-fat diet increases the activity of the FAS enzyme (WAKIL, 1989; AILHAUD, 2008).

High levels of MUFA in adipose tissue and liver may be linked to stimulation of the activity of Stearoyl-CoA desaturase, a key enzyme in the synthesis of MUFA. In adipose tissue and the liver, stearic acid (C18:0), (known to be the preferred substrate for delta-9 desaturase), undergoes delta-9 desaturation to give oleic acid (18:1n-9). In the adipose tissue and liver of obese rats, stearic acid is converted to oleic acid to a much greater extent than in control rats. This increase may be the result of greater delta-9 desaturase activity in the adipose tissue and liver, but it may also be explained by a different need for quality GAs in these tissues, since adipose tissue has a role in storing GAs for the body as a whole, while the liver uses GAs to synthesise lipoproteins or oxidises them for energy purposes. It has been shown that the activity of the Stearoyl-CoA desaturase enzyme in adipocytes is high in the aged obese Zucker rat **(JONES et al., 1996).** Several authors have noted, in obese Zucker rats, a reduction in C20:4(n-6) compared with C18:2(n-6) both in tissues and in circulating lipids, which is explained by an alteration in the metabolism of PUFAs. **Some authors have noted an alteration in PUFA metabolism, with an increase in MUFA and a decrease in long-chain polyunsaturated fatty acids in rats fed a cafeteria diet** (LLADO et al., 1996). **Obesity is associated with** an increase in delta 9 desaturase and a reduction in delta 5 desaturase and delta 6 desaturase following **insulin resistance (WAKIL and ABU-ELHEIGA, 2009).** These observations support our results. The increase in C18:2(n-6) in the adipose tissue and liver of elderly obese rats may be due to inhibition of the conversion of C18:2(n-6) to C20:4(n-6), and the reduction in long-chain PUFAs (C20:4n-6; C20:5n-3 and C22:6n3) following their reduced synthesis, an obvious consequence of the decrease in the activity of delta 5 and delta 6 desaturases. In general, these changes may therefore be linked to the effect of the cafeteria diet, changes in desaturase activity and insulin resistance.

Several authors have established that excess body fat results in dyslipidaemia characterised by high concentrations of total cholesterol, triglycerides, low-density lipoproteins and apolipoprotein B, and low concentrations of high-density lipoproteins and apoprotein A (CAPRIO et al., 1996; FREEDMAN et al., 1999; TEIXEIRA et al., 2001).

The experimental rats had higher protein levels in the LDL and HDL fractions compared with their controls, and this increase was probably due to the increase in apoprotein B. On the other hand, our results reveal an increase in total protein content in serum and organs (liver and adipose tissue) in experimental rats compared with control rats. Indeed, the maintenance of body protein mass results from the balance between protein synthesis and catabolism according to a rhythm dependent on exogenous nitrogen inputs (LACOIX et al., 2004). This increase may be due to the 'cafeteria' diet or to ageing, which includes disturbances in the regulation of protein metabolism (BOIRE et al., 2005).

Creatinine and urea are excellent markers of renal function, and their increase or decrease reflects renal dysfunction. Creatinine is the best marker of renal function. It is formed in the body by the non-enzymatic dehydration of creatine synthesised by the liver and stored in the muscle. A high serum creatinine level (associated with a high urea level) indicates a reduction in glomerular filtration. Urea, another marker of renal function, is produced by the destruction of proteins. It is completely filtered by the glomeruli. Its blood level reflects the overall functioning of the kidneys.

Urea and creatinine levels were elevated in obese rats compared with controls, possibly due to abnormal renal function caused by the cafeteria diet or changes in metabolism during ageing, which is characterised by a gradual decline in biological functions caused by progressive dysfunction of the various cellular systems involved in repairing and maintaining homeostasis. As a result, irreversible damage accumulates in lipids, proteins and nucleic acids (HOLLIDAY, 2006; RATTAN, 2008).

Disruptions to the antioxidant system have been reported in obesity and ageing (FURUKAWA et al., 2004; AUGUSTYNIAK et al., 2005; SENTHIL KUMARAN et al., 2008; KUMARAN et al., 2009). Oxidative stress occurs when the production of free radicals exceeds the capacity of the antioxidant defence system. Free radicals cause damage to cells, lipids and proteins, leading to various pathologies (HIGDON and FREI, 2003; MERZOUK S et al., 2004).

Obesity and ageing increase the rate of lipid peroxidation and protein carbonylation (FURUKAWA et al., 2004; DELATTRE et al., 2005). Measuring the levels hydroperoxides, MDA and carbonylated proteins is a determining factor in oxidant/antioxidant status.

In our work, the results obtained support the existence of oxidative stress in aged rats and also in obese rats (KREGEL, 2007; BOUANANE et al., 2009). Elderly obese rats fed a cafeteria diet show an increase in plasma HYDR and plasma and tissue malondialdehyde (MDA) and carbonylated protein (PCAR) levels, indicating obvious oxidative stress. It has been shown that a high-calorie, high-fat diet increases the production of free radicals and reduces antioxidant defence capacity (SREEKUMAR et al., 2002; MILAGRO et al., 2006). In addition, in obesity, oxidative stress may be generated as a result of the oxidation of excess nutrients (UNGER, 2003). It should be remembered that these rats are obese rats with increased food intake and excess adipose tissue and are also aged rats, and numerous studies have highlighted the involvement of oxidative stress in ageing (ROMANO and SERVIDDIO., 2010).

Lipid peroxidation has been estimated by measuring HYDPs, the essential products of lipid peroxidation, polyunsaturated fatty acids (PUFAs) or their esters (e.g. phospholipids and triglycerides) (MICHEL et al., 2008). Lipid peroxidation leads to the release of other oxidation products such as conjugated dienes and aldehydes which, in high concentrations, are toxic to cells. Most of these aldehydes are highly reactive and can be considered toxic second messengers that increase the initial damage caused by free radicals. The best-studied aldehyde is MDA, formed during the cleavage of polyunsaturated fatty acids with at least three double bonds (ESTERBAUER et al., 1989). The increase in HYDP in plasma and MDA in plasma, liver, muscle and adipose tissue in elderly obese rats fed a cafeteria diet is in favour of increased lipid oxidation.

The in vitro oxidation of plasma lipoproteins, induced by metals (copper), is determined by monitoring the formation of these conjugated dienes over time. The formation of conjugated dienes results from the rearrangement of the ethylenic double bonds of polyunsaturated fatty acids (PUFAs) following the radical abstraction of a malonic hydrogen (ESTERBAUER et al., 1989). Conjugated dienes are higher in elderly obese rats compared with controls, suggesting increased formation of lipid peroxidation products. This result may be interpreted as a very rapid oxidation of lipoproteins over time. Some authors have noted the rapid oxidation of lipids that occurs in the LDL of obese people as a result of the reduction in antioxidants (CRUJEIRAS et al., 2006). In fact, the concentration of vitamin E, which inhibits the propagation of chain reactions reacting with free radicals, and vitamin C, which prevents the oxidation of LDL and regenerates oxidised vitamin E, are reduced in obese people and even in the elderly (CRUJEIRAS et al., 2006 ; KREGEL , 2007).

Given that lipoproteins are sensitive to oxidation phenomena, we can assume that the lipoproteins of obese rats are less resistant to oxidation in vitro compared with those of controls (ESTERBAUER et al., 1989). Our results are in agreement with those of KELISHADI et al (2007), VINCENT et al (2007), and UZUN et al (2007) who show that obesity increases oxidative stress by increasing lipoprotein oxidation.

Carbonylated proteins are considered to be markers of protein oxidation. A significant increase in plasma and tissue carbonylated protein levels was observed in elderly obese rats compared with controls. These data are in agreement with those of VINCENT et al (2007) and UZUN et al (2007) who show that carbonylated protein levels increase in obese people, and with those of DELATTRE et al (2005) who show that carbonylated protein levels increase in elderly people. Protein oxidation is a sign of tissue damage caused by oxidative stress, increased carbohydrate levels, or both (KREGEL, 2007). It should be remembered that elderly obese rats have hyperglycaemia, which can induce glycation and protein oxidation.

Several authors have reported that the cafeteria diet induces an increase in the formation of free radicals following an alteration in mitochondrial oxidative mechanisms, associated with an increase in lipid peroxidation, lipoprotein oxidation and protein oxidation (SREEKUMAR et al., 2002; LOPEZ et al., 2003; MILAGRO et al., 2006; GARCIA-DIAZ et al., 2007). In our work, the cafeteria diet not only causes excessive production of free radicals, but also reduces the antioxidant defence capacity by reducing certain antioxidant enzyme activities in elderly obese rats.

The body has a complex antioxidant system, including both enzymatic and non-enzymatic components, which protects biomolecules (proteins, lipids, etc.) against the harmful effects of free radicals. The hyperproduction of free radicals, and therefore the tissue damage they cause, is limited by the natural endogenous presence of antioxidant substances. Other systems for destroying free radicals are not enzymatic but strechiometric, with the molecules reacting one by one; once they have reacted with a free radical, they are destroyed. The main strechiometric destroyer of free radicals is a-tocopherol (vitamin E), which inhibits the propagation of the oxidative chain by reacting with free radicals. In addition to its antioxidant role, ascorbic acid (vitamin C) regenerates vitamin E. Vitamin A inhibits lipid peroxidation, but can also directly inhibit hydroxyl radicals (JAESCHKE, 1995). As regards markers of antioxidant defence, in this study we measured plasma vitamin C levels at . We also measured the antioxidant activity of catalase, which catalyses the dismutation of hydrogen peroxide into water and molecular oxygen, and glutathione, a tripeptide formed by the condensation of glutamic acid, cysteine and glycine: γ-L-Glutamyl-L- cysteinylglycine. Glutathione, which exists in oxidised and reduced forms, is involved in maintaining the redox potential of the cell's cytoplasm. It is also involved in a number of detoxification reactions and the elimination of reactive oxygen species. Note that the amine group of cysteine condenses with the γ-carboxylic acid function of glutamic acid. Virtually all cells contain a high concentration of cysteine. It is simplified by GSH (reduced form) or GSSG (oxidised form), the thiol function giving it its main biochemical properties.

Our results show that plasma vitamin C levels are reduced in elderly obese rats compared with control rats. These results are in line with those obtained in humans by SINGH et al (1994; 1998) who showed a significant positive association between obesity and a reduction in serum beta-carotene and antioxidant vitamin C and E levels. In addition, Perticone et al (2001) indicate a decrease in serum vitamin C levels during human obesity. The activities of antioxidant enzymes are also modified in obese rats. Catalase and GSH activities are reduced in plasma, but GSH is increased in liver and adipose tissue in aged obese rats. Several studies have reported different antioxidant enzyme activities during obesity and ageing, indicating either an increase or a decrease (VINCENT and TAYLOR, 2006; JAYAKUMAR et al., 2007; SENTHIL KUMARAN et al., 2008; KUMARAN et al., 2009; AYDIN et al., 2010). Indeed, in the face of oxidative stress, antioxidant enzymes are consumed and inactivated, which may explain the reduction in plasma catalase and GSH activities in elderly obese rats.

RAJASEKARAN et al (2002) show that the decrease in GSH is the cause of the decrease in ascorbic acid and a-tocopherol. In addition, the formation of free radicals stimulates and activates antioxidant defence, which may explain the increase in GSH activity in the liver and adipose tissue of elderly obese rats (DELATTRE et al., 2005). No difference was observed in muscle and intestine.

Taken together, the results suggest that obesity and ageing are associated with changes in lipid and lipoprotein metabolism and oxidant/antioxidant status. It is therefore necessary to look for simple means, particularly nutritional, to correct these metabolic anomalies. The second objective of our work is therefore to study the role of (n-3) PUFAs provided by flaxseed oil, which are lipid-lowering agents that are also thought to have anti-atherogenic and antioxidant properties.

The study by BURR et al (1989) demonstrated the benefits of n-3 PUFAs by administering 3 different diets to 3 groups suffering from coronary heart disease. The first group received a diet with a moderately high intake of linoleic acid, the second group a diet rich in fish and the third a diet rich in fibre, all with a low intake of saturated fats. Only the fish group had a 29% reduction in coronary and total mortality.

Epidemiological studies have also shown a reduction in the incidence of inflammatory pathologies in Japanese and Greenland Eskimos. This is attributed to a high consumption of cold sea fish rich in n-3 PUFAs (JOLLY et al., 1997 ; MILES and CALDER, 1998 .WESLY et al., 1998).

Our results show that the diet enriched with flaxseed oil leads to a reduction of hyperglycaemia with a marked decrease with the percentage of 5% flaxseed oil. This is in agreement with the results of FASHING et al (1991) who reported that the administration of fish oil for two weeks to obese patients suffering from insulin resistance significantly increased their insulin sensitivity. Our results also show that supplementing the diet with flaxseed oil resulted in a slight weight loss in elderly control rats. However, obese elderly rats fed the cafeteria diet supplemented with linseed oil had a significantly lower body weight than obese elderly rats fed the cafeteria diet alone, and effect of linseed oil at 5% was more marked than at 2.5%. Our results are in agreement with those of FICKOVA et al (1998) who showed that rats fed a diet enriched with PUFAIn-3 had a lower body weight and insulin concentration than rats fed a diet enriched with n-6 PUFA. Studies in animals and humans have shown that polyunsaturated fatty acids (PUFAs) are more easily used as fuel, while saturated fatty acids (SFAs) are more likely to be accumulated in adipose tissue (HARIRI et al., 2010). According to (SHIROUCHI et al., 2007), this loss of fat mass is induced by a reduction in the proliferation of preadipocytes and adiposity. Also, n-3 PUFAs reduce adipocyte lipogenesis by increasing the sensitivity of muscle tissue to insulin and therefore reducing glucose transport, as well as reducing lipid storage in adipose tissue (LOVEJOY, 1999). JUCKER et al (1999) compared the effects of consuming n-3 PUFAs with the effects of consuming n-6 PUFAs, and concluded that the latter increase intramuscular triglyceride levels, lead to insulin resistance and reduce muscle glycolysis. This is why the n-3/n-6 balance is an important factor in cellular metabolism. OAKES et al (1997) have attributed the increase in intramuscular triglyceride levels to the in situ reduction in insulin levels responsible for stimulating glucose metabolism in the muscle and glycogenesis in the liver. n-3 PUFAs also reduce hyperlipidaemia and the long-term complications of obesity. HARRIS et al (,1988), CONNOR (1986), SIRTORI et al (2002) have shown a reduction in serum TG and TC levels under the effect of n-3 PUFAs. Our results are in line with those of previous studies.

We noted that linseed oil corrected the lipid disorders of elderly obese rats. This is represented by a reduction in serum TG levels resulting from a reduction in hepatic, adipocyte and muscle TG levels.

Several studies have shown that the reduction in TG levels in insulin-dependent tissues is at the origin of the reduction in serum TG levels (WONG et al., 1985; HOORROCKS and YEO, 1999). In our study, we also observed that the flaxseed oil-enriched diet had a cholesterol-lowering effect on the lipoproteins and serum of elderly obese rats. The cholesterol-lowering effect of the n-3 PUFAs contained in linseed oil was accompanied by a reduction in hepatic and adipose tissue cholesterol, suggesting a reduction in cholesterol synthesis or an increase in its excretion into the bile. In fact, it has been shown that fish oil induces variations in cholesterol metabolism in the liver of rats, leading to an increase in its excretion into the bile (CONNOR, 1986).

Several studies have also reported the cholesterol-lowering effect of n-3 PUFAs in rats (MERZOUK and KHAN, 2003; SOULIMANE et al., 2005). However, in humans, the effect of PUFAs on cholesterol metabolism is debated. SANDERS et al (1983) reported that 20g/day of fish oil (containing 5g of n-3 PUFAs) significantly reduced serum TG levels in humans, whereas cholesterol levels were only reduced in subjects with high initial cholesterol

levels.
However. In our study, it is important to emphasise that the flaxseed oil-enriched diet resulted in a significant increase in HDL-C in obese elderly rats. These results are consistent with those of SANDERS et al (1983). HDL is known for its anti-atherogenic effect and its protection of VLDL against oxidative modifications (HIGDON AND FREI, 2003).
Our results show a reduction in proteins in the LDL and HDL fractions, as well as a reduction in adipose tissue and liver, whereas no difference was noted in muscle and intestine in rats on a cafeteria diet enriched with linseed oil. FAILOR et al (1988) noted a significant reduction in apoproteins B in subjects consuming In-3 PUFAs.
GRUNDY (2006) explained the effect of n-3 PUFAs on the reduction of LDL and hypercholesterolaemia by facilitating the clearance of chylomicrons and by reducing competition from small quantities of VLDL. Generally speaking, the lipid-lowering effect of n-3 PUFAs can be explained as follows:
These fatty acids can reduce TG synthesis and chylomicron secretion by intestinal cells (HARRIS, 1989) and suppress hepatic synthesis of fatty acids and TG production, thereby limiting VLDL secretion (CONNOR, 1985; 1986; WONG and MARSH, 1988). In rats, n-3 PUFAs reduce the activity of glucose-6-phosphate dehydrogenase, the malic enzyme and acetyl-CoA carboxylase (IRITANI et al., 1980). EPA reduces hepatic TGs by inhibiting phosphatidate phosphohydrolase and acyltransferases (RUSTAN and DREVON, 1989; WONG and MARSH, 1988). n-3 PUFAs reduce the TG pool in VLDL by inhibiting TG synthesis and possibly by increasing the clearance of TG-VLDL (WONG and MARSH, 1988; HARRIS, 1989). These fatty acids also reduce the synthesis of apoproteins (ILLINGWORTH et al., 1984; WONG and MARS, 1988). It has been reported that EPA reduces the production of apoproteins in humans and rats (NESTELET al., 1984; 1986) and can reduce VLDL levels.
Our results also show a decrease in LCAT activity in rats fed a diet enriched with linseed oil. These results are in agreement with those of TSUKAMOTO et al (1982) and SUBBAIAH et al (1998) who found that consumption of a diet rich in PUFAs reduced LCAT activity. SINGER et al (1990). found the same thing after a diet rich in linseed oil (rich in n-3 PUFAs).
The results of LI et al (2005) showed an increase in bile acids or their excretion in mice fed a diet enriched with PUFAs, which simultaneously reduced the accumulation of cholesterol in the liver. The effect of n-3 PUFAs on LPL activity is debated; some authors have found that this activity remains unchanged on a diet enriched in n-3 PUFAs (DAVID et al., 1987). However, our results show an increase in adipocyte and hepatic LPL activity in rats on a diet enriched with linseed oil. This is not in line with the results of HAUG and HOSTMARK . (1987) who noted a 50% reduction in LPL activity in rats consuming a diet enriched with fish oil.
creatinine and urea levels in rats on a diet enriched with linseed oil showed a reduction compared with those in rats on a cafeteria diet alone, which may be in favour of normal renal function in aged rats and also of a beneficial effect of linseed oil on renal function (KELLEY et al., 1991).
DAS et al (2001) suggest that EPA and DHA supplementation inhibits the production of free radicals and suppresses lipid peroxidation in patients suffering from nephritic syndrome. These results suggest that EPA or DHA may be involved in the elimination of free radicals.
Our results show that the composition of FAs in the serum of the liver and adipose tissue of aged rats is modified after enrichment of the diet with linseed oil, with a decrease in SFAs, MUFAs and C18:2n-6 and an increase in PUFAs, C18:3n-3, C20:5n-3 and C20:4n-6,

reflecting the difference in the fatty acids that make up the diet and the improvement in desaturase activities. Our results are in line with those of (MIRET et al.,2003 ; HUSSEIN et al .,2007)

Our results also indicate that linseed oil increases vitamin C levels and the activities of erythrocyte antioxidant enzymes. On the other hand, it reduces the levels of hydroperoxides and carbonylated proteins in obese elderly rats, which is consistent with several studies (KESAVULU et al., 2002; SARSILMAZ et al., 2003; JAYAKUMAR et al., 2007; SENTHIL KUMARAN et al., 2008). The flaxseed oil-enriched diet improves oxidant/antioxidant status and reduces oxidative stress induced by obesity and age, which highlights the beneficial effects of n-3 PUFAs and is consistent with the results of YESSOUFOU et al (2006).

The reduction in tissue susceptibility to oxidation by n-3 PUFAs may be linked to an improvement in cell membrane stability, after incorporation of these fatty acids into plasma membrane phospholipids (YUAN and KITTS, 2002). BRUDE et al (1997) reported that the treatment of obese patients with n-3 PUFAs may be linked to an improvement in the membrane sensitivity of cells, after incorporation of these fatty acids into the phospholipids of the plasma membrane (YUAN and KIETTS, 2002). BRUDE et al (1997) reported that the accumulation of EPA and DHA in the plasma membrane reduces the attack on double bonds by free radicals or H_2O_2.

The fact that linseed oil is rich in α-linolenic acid, which after elongation and desaturation gives eicosapentaenoic acid (EPA), may explain why this fatty acid can make a major contribution to antioxidant properties. DEMOZ et al (1992) and VENKATRAMAN et al (1994) have reported that In-3 PUFAs modulate the activities of antioxidant enzymes and increase the effectiveness of the body's antioxidant defence system. It has also been shown that the fatty acids in fish oil adhere to the phospholipid composition of the cell membrane, resulting in increased levels of EPA and DHA at the expense of arachidonic acid (ANDO et al., 1998). This substitution may reduce negative effect of arachidonic acid (n-6) on antioxidant status. In addition, the fatty acid composition of the membrane influences the physical properties of the cell membrane (fluidity, permeability), the activity of receptors, enzymes and ion channels, as well as the cellular response of second messengers to various stimuli.

The beneficial effect of In-3 PUFAs involves EPA and DHA. It is known that EPA increases the eicosanoids of the n-3 family, which exert an opposite effect to the eicosanoids of the n-6 family (derived from arachidonic acid). EPA can also be converted into DHA, which in addition to dietary DHA, can contribute further to the beneficial effects. One study showed that DHA can increase certain derivatives, such as docosatrienes or resolvins, which also exert a beneficial effect (SHERHAN et al., 2004). However, the mechanism by which EPA and DHA exert a beneficial effect on antioxidant status has not yet been demonstrated.

In addition, fish oil increases the activity and mRNA levels of catalase, glutathione peroxidase, superoxide dismutase (VENKATRAMAN et al., 1994) and liver glutathione reductase in mice (DEMOZ et al., 1992).

KHAN and HICHAMI. (2002) have also shown that In-3 PUFAs modulate cell signalling mechanisms via their incorporation into cell membrane phospholipids.

In an aqueous medium, the methyl ester of EPA and linoleate is present in micellar form, and the oxidability of EPA methyl ester is lower than that of linoleate methyl ester. EPA micelles have two or more oxygen molecules in their peroxyl radical, while linoleate micelles have only one molecule (YAZU et al., 1998). These authors discussed the fact that EPA is more polar than linoleate, and that polar radicals can migrate from the lipophilic core of the micelle to the polar surface. This migration creates an environment favourable to the formation of stable compounds, which reduces the propagation of oxidative reactions (WANDER and SHI-HUA, 2000). This is one of the mechanisms that may explain the antioxidant properties of EPA, which acts as a peroxyl and a scavenger of free radicals. Only the mechanisms by which DHA acts as an antioxidant have not yet been elucidated.

These results clearly show that n-3 PUFAs have a beneficial effect on lipid metabolism disorders and oxidant/antioxidant status, which are altered by obesity and age.

CONCLUSION

Obesity has become a major public health problem, with this epidemic preceding an impressive wave of pathologies (type II diabetes, dyslipidemia, cardiovascular disorders....). The World Health Organisation has made its prevention and management a priority in the field of nutritional pathology.

Furthermore, ageing, which is characterised by a slowing down of all metabolic pathways, as well as alterations in carbohydrate, lipid and protein homeostasis and intense oxidative stress, provides a favourable breeding ground for the development of numerous diseases.

The combination of obesity and ageing can therefore aggravate metabolic alterations.

In the context of the fight against obesity and ageing and the reduction of their progression, nutritional prevention through diets enriched with n-3 PUFA-rich oil has a special place. In our study, we sought to determine the effects of the cafeteria diet (hypercaloric and hyperlipidic) and flaxseed oil supplementation (2.5% and 5%) on body weight and metabolic changes (lipid and protein parameters and oxidant/antioxidant status) using an experimental model of nutritional obesity, aged wistar rats.

Epidemiological data suggest that a high-fat diet favours the development of obesity, so we used an experimental model of nutritional obesity: the 'cafeteria' diet. In this model, animals are offered several types of highly palatable, energy-dense human food in order to induce voluntary hyperphagia, which leads to the onset of obesity. The use of this diet in Wistar rats enables us to demonstrate the onset of obesity.

Our results show that the "cafeteria" diet induces a significant increase in body weight in aged rats. This excess weight is associated with an accumulation of adipose tissue and its enrichment in lipids (increased TC and TG). Hyperlipidaemia was also observed in the liver and adipose tissue. In addition, the lipid profile of aged rats consuming the 'cafeteria' diet was characterised by an increase in serum and lipoprotein TC and TG levels and a decrease in HDL-C. There was also a reduction in LCAT activity, which may explain the drop in HDL-C levels, as well as an increase in MUFA and C18:2(n-6).

Our results also show an increase in LPL activity in the adipose tissue and liver of obese rats on the cafeteria diet, which enables lipids to be captured and stored in these organs. The over-expression of LPL activity may be linked to a state of insulin resistance, due to the hyperglycaemia observed in these rats consuming the 'cafeteria' diet, as well as the increased lipolytic activity of LHS in the adipose tissue of these same rats, where insulin has little influence on lipolysis. It is clear a high-fat, high-calorie diet leads to increased accumulation of lipids in adipose tissue, resulting in obesity that is strongly associated with insulin resistance and ectopic accumulation of lipids in other organs such as the liver, resulting in metabolic disorders that can lead to cardiovascular events.

In terms of oxidant/antioxidant status, the results obtained reflect significant oxidative stress. In these elderly obese rats, we noted an increase in plasma and tissue levels of MDA, PCAR and markers of lipoprotein oxidation and the activity of catalase and glutathione in plasma, and a decrease in vitamin C.

Flaxseed oil supplementation in the cafeteria diet confirms the beneficial influence of n-3 PUFAs on body weight, with a reduction in lipogenesis. At organ level, linseed oil significantly reduces the average weight of adipose tissue in obese people. In addition, linseed oil reduced glycaemia in control rats and plasma and tissue lipids (liver, muscle, intestine, adipose tissue), confirming the lipid-lowering and cholesterol-lowering effect of linseed oil.

This effect was very marked in obese age-group rats (CAFL5%).
Our results also show the beneficial effect of linseed oil on the production of oxidised markers, leading to a reduction in the production of MDA, HYDP and PCAR, and acting on markers of in vitro oxidation of plasma lipoproteins by limiting the rate of lipoprotein oxidation. Linseed oil also acts on organs by limiting the production of hepatic, muscular and adipose tissue MDA and PCAR. The beneficial effect of linseed oil on the tissue redox profile appears to be more marked in obese people compared with controls. Through its n-3 PUFAs, linseed oil modulates the activity of antioxidant enzymes and therefore regulates markers of oxidative status in aged rats.
This study led us to conclude that flaxseed oil (2.5% and 5%) has a beneficial effect on reducing obesity and the consequences of ageing by regulating metabolic parameters and the redox balance.
Following on from our work, we want to use this same line of research to extend our knowledge of the mechanisms of action of flax on the pathophysiology of elderly obese rats, using molecular investigation .

BIBLIOGRAPHICAL REFERENCES

1. ABDEL-MONEIM AE, DKHIL MA, AL-QURAISHY S (2010) The Redox Status in Rats Treated with Flaxseed Oil and Lead-Induced Hepatotoxicity. Biol Trace Elem Res. 2010 Oct 20 (Epub ahead of print).

2. AEBI H (1974) Catalase In methods of enzymatic analysis. Verlag Chimie Gmbh Weinheim. 2: 673 - 684.

3. AFSSA (2001). Acides gras de la famille oméga 3 et système cardiovasculaire : intérêts nutritionnel s. http://www.mangerbouger.fr/pro/IMG/pdf/AcidesGrasAfssa.pdf.

4. AILHAUD G (2002) Autocrine/paracrine effectors of adipogenesis. Ann. Endocrinol (Paris) .63:83-5

5. AILHAUD G (2008) Apports lipidiques et prise de poids : aspects qualitatifs. OCL. 15: 37-40.

6. AILHAUD G (2007) Développement du tissu adipeux : importance des lipides alimentaires. Centre de Biochimie .UNSA .Nice : 4-6.

7. AILHAUD G , GUESNET P (2003) Fatty acid composition of fats in an early determinant of obesity: a short review and opinion.Obesity reviews (s press).

8. ALBERS JJ, CHEN CH, LACKO AG (1986) Isolation, characterization, and assay of lecithin-cholesterol acyltransferase. Methods Enzymol. 129: 763-783.

9. ALBERTI K G, ZIMMET P, SHAW J (2005) The metabolic syndrome-a new worldwide definition. Lancet 366(9491): 1059-62.

10. ANDO K, NAGATA K, BEPPU M, KIKUGAWA K, KAWABATA T, HASEGAWA K, SUZUKI M (1998) Effect of n-3 fatty acid supplementation on lipid peroxidation and protein aggregation in rat erythrocyte membrane. Lipids.33:505-512.

11. ANDRE C (2008) Etude du rôle des cytokines dans l'activation de l'indoléamine 2,3-dioxygénase cérébrale impliquée dans les altérations comportementales associées à l'inflammation. [Doctoral thesis in life and health sciences]: Université Toulouse III.

12. ARTHUR Y, HERBETH B, GUENOURI L, LECOMTE E, JEANDEL C, SIEST G (1992) Age-related variations of enzymatic defenses against free radicals and peroxides. In: Emerit I, Chance B, eds. Free radicals and aging. Basel: Birkhaüser Verlag.359-67.

13. AUBIN MC (2009) Étude de la fonction vasculaire et du remodelage cardiaque avant l'établissement de l'obésité et de la dyslipidémie chez les rats femelles Sprague-Dawley recevant une régime riche en gras. [D. thesis in Pharmacology]: Université de Montréal, Faculté de Médecine.

14. AUGUSTYNIAK A, WASZKIEWICZ E, SKRZYDLEWSKA E (2005) Preventive action of green tea from changes in the liver antioxidant abilities of different aged rats intoxicated with ethanol. Nutrition .21:925-932.

15. AYDIN A.F, KUCUKGERGIN C, OZDEMIRLER-ERATA G, KOCAK-TOKER N, UYSAL M (2010) The effect of carnosine treatment on prooxidant-antioxidant balance in liver, heart and brain tissues of male aged rats. Biogerontology .11: 103-109.

16. BARANOWSKI M, ENNS J, BLEWETT H, YAKANDAWALA U, ZAHRADKA P, TAYLOR CG (2012) Dietary flaxseed oil reduces adipocyte size, adipose monocyte chemoattractant protein-1 levels and T-cell infiltration in obese, insulin-resistant rats. Cytokine .59:382-91.

17. BARBER T, VINA JT, CABO J (1985) Decreased urea synthesis in cafeteria-diet-induced obesity in the rat. Biochem J. 230: 675-681.

18. BASDEVANT A (2006) L'obésité: origines et conséquences d'une épidémie. C. R. Biologies.329: 562-569.
19. BASDEVANT A, GUY-GRAND B (2004) Traité de médecine de l'obésité, Flammarion Médecine Sciences, Paris.
20. BAUER JD (1982) Clinical laboratory methods. St. Louis, C.V. Mosby Co, 9th ed
21. BEAUDEUX JL, VASSON MP (2005) Sources cellulaires des especes réactives de l'oxygénes in: Radicaux libres et stress oxydant Aspects bilogiques et pathologiques. Coordinators DELATTRE J, BEAUDEUX JL, BONNEFONT-ROUSSELOT D. Editions Lavoisier, Paris: 45-86.
22. BENKALFAT N, MERZOUK H, BOUANANE S, MERZOUK SA, BELLENGER J, GRESTI J, TESSIER C, NARCE M (2011) Altered adipose tissue metabolism in offspring of dietary obese rat dams. Clinical science. 121(1):19-28
23. BENNETT AJ, BILLETT MA , SALTER AM, WHITE DA (1995) Re gulation of ham s ter mi crosom al triglyceride transfer protein m R NA le vels by dietary fats. Biochem. Biophys. Res. Commun. 212 : 473-478.
24. BERR C, NICOLE A, GODIN J (1993) Selenium and oxygen-metabolizing enzymes in elderly commuity residents. A pilot epidemiological study. J Am Geriatr Soc .41 : 143-8.
25. BESANÇON P (2001) Besoins alimentaires et qualité nutritionnelle des aliments. In. J.Cheftel. H. Cheftel. P. Besançon. Technique et documentation. Paris. Lavoisier. 5: 89 - 134.
26. BESSESEN DH, ROBERTSON AD, ECKEL RH (1991) Weight reduction increases adipose but decreases car Diabete sc LPL in reduced-obese Zucker rats. Am J Physiol.261: E246-E251.
27. BJOMTORP P (1991) Metabolic implication of body fat distribution. Diabetes.14: 11321143.
28. BLIGH EG, DYER WJ (1959) a rapid method of total lipid extraction and purification. Can.J.Physiol.Pharmacol.37:911-917.
29. BLUNDELL JE, KING NA (1996) Overconsumption as a cause of weight gain: behavioural physiological interactions in the control of food intake (appetite). In: Chadwick DJ, Cardew GC. The origins and consequences of obesity. Chichester (UK), Wiley: 138-158.
30. BODEN G, SHULMAN GI (2002) Free fatty acids in obesity and type 2 diabetes: defining their role in the development of insulin resistance and beta-cell dysfunction. Eur J Clin Invest. 32 (Suppl 3):14-23.
31. BOIRIE Y, GUILLET C, ZANGARELLI A, GRYSON C., WELTROND S (2005) Altérations du métabolisme protéique au cours du vieillissement. Clinical Nutrition and Metabolism. 19(3):138-142.
32. BOKOV A, CHAUDHURI A, RICHARDSON A (2004) The role of oxidative damage and stress in aging. Mech Ageing Dev.125: 811-26.
33. BONNEFONT R, BEAUDEUX JL, DELATTRE J (2003) Radicaux libres et stress oxydant. Biological and pathological aspects. Lavoisier. Edition. DOC. Editions Médicales. Internationales. Paris. P : 147 - 167.
34. BOOZER CN, BRASSEUR A, ATKINSON RL (1998) Dietary fat affects weight loss and adiposity during energy restriction. Am J Clin Nutr. 58: 846-842.
35. BOUANANE S, BENKALFAT NB, BABA AHMED FZ, MERZOUK H, MOKHTARI NS, MERZOUK SA, GRESTI J, TESSIER C, NARCE M (2009) Time course of changes in serum oxidant/antioxidant status in overfed obese rats and their offspring. Clinical Science. 116:669-680.

36. BOUHALI T (2006) L'adiponectine, un modulateur du risque de maladie coronarienne athérosclérotique dans l'hypercholestérolémie familiale. [Doctoral thesis in life and health sciences]: Université Laval Québec, Faculté de médecine.
37. BRANCA F (2008) Opening Ceremony. 16th European Congress on Obesity, WHO. Obesity is a critical public health problem that affects many lives, many communities and many nations. 16th European Congress on Obesity-ADGSpeech.pdf.
38. BRAUN JE, SEVERSON DL (1992) Regulation of the synthesis, processing and translocation of lipoprotein lipase. Bi ochem. J. 287: 337-347.
39. BRUDE IR, DREVON C, HJERMANN I, SELJEFLOT I, LEND-KATZ S, SAAREM K, SANDSTAD B, SOLVOL K, HALVORSEN B, ANESEN H, NENSETER M S (1997) Peroxidation of LDL from combined- Hyperlipidemic male smokers supplied with omega-3 fatty acids and antioxidants. Arterio. Thromb. Vasc.Biol.17:2576-2588.
40. BUCKLEY JD, HOWE PR (2010) Long-chain omega-3 polyunsaturated fatty acids may be beneficial for reducing obesity-a review. Nutrients.2:1212-30.
41. BUETTNER GR (1993) the pecking order of free radicals and antioxidants : lipid peroxidation, alpha-tocopherol, and ascorbate . Arch.Biochem. Biophys.300 (2):535-543.
42. BUETTNER R, PARHOFER KG, WOENCKHAUS M, WREDE CE, KUNZ-SCHUGHART LA, SCHÖLMERICH J, BOLLHEIMER LC (2006) Defining high-fat- diet rat models: metabolic and molecular effects of different fat types. Journal of Molecular Endocrinology. 36: 485-501.
43. BURR M, FEHILY A.M, GILBERT J.F, ROGERS S, HOLLIDAY R.M, SWEETNAM P M, ELWOOD, PC, DEADMAN N.M (1989) Effects of changes in fat fish and fibre intakes on death and myccardial reinfaction: diet and reinfaction trial (DART). Lancet.334: 757-761.
44. BURSTEIN M, FINE A, ATGER V (1989) Rapid method for the isolation of two purified subfractions of high density lipoproteins by differential dextran sulfate magnesium chloride precipitation. Biochem. 71:741-746.
45. BURSTEIN M, SCHOLNICK HR, MORFIN R (1970) Rapid method for the isolation of lipoproteins by precipitation with polganions. JLR. 11: 583-595.
46. CABALLERO B (2007) The global epidemic of obesity: an overview. Epidemiol Rev. 29:1-5.
47. CADET J, BELLON S, BERGER M, BOURDAT A.G, DOUKI T, DUARTE V, FRELON S, GASPARUTTO D, MULLER E, RAVANAT J.L, SAUVAIGO S (2002) Recent aspects of oxidative DNA damage: guanine lesions, measurement and substrate specificity of DNA repair glycosylases. Biol. Chem. 383(6): 93.
48. CAPRIO S, HYMAN LD, MCCARTHY S, LANGE R, BRONSON M, TAMBORLANE WV (1996) Fat distribution and cardiovascular risk factors in obese adolescent girls: importance of the intra abdominal fat depot.Am J Clin Nutr. 64(1):12-7.
49. CARLSON LA, HALLBERG D, MICHELI H (1969) Quantitative studies on the lipolytic response of human subcutaneous and omental adipose tissue to noradrenaline and theophylline. Acta Med Scand. 185(6):465-9.
50. CARROLL L, VOISEY J, VAN DAAL A (2004) Mouse models of obesity. Clinics in dermatology. 22: 345-349.
51. CARTER J (1993) Flax seed as a source of alpha linolenic acid. Journal of the American College of Nutrition. 12(5):551.
52. CEBALLOS-PICOT I, TRIVIER JM, NICOLE A, SINET PM, THEVENIN M (1992) Age-corrated modifications of copper-zinc dismutase and glutathione-related enzyme

activities in human erythrocytes. Clin Chem.38 : 66-70.

53. CHAPMAN J (1982) Lipoproteins and the liver. Gastro enterol Clin. Biol. 6 : 482-499.

54. CHAPMAN MJ, SPOSITO AC (2008) Hypertension and dyslipidaemia in obesity and insulin resistance: pathophysiology, impact on atherosclerotic disease and pharmacotherapy. Pharmacology & Therapeutics. 117(3):354-373.

55. CHICCO AG, D'ALESSANDRO ME, HEIN GJ, OLIVA ME, LOMBARDO YB (2009) Dietary chia seed (Salvia hispanica L.) rich in a-linolenic acid improves adiposity and normalises hypertriacylglycerolaemia and insulin resistance in dyslipaemic rats. Brit J Nutr. 101:41-50.

56. CONNOR W.E, BRISTOW JD (1985) Coronary heart desease prevention complications and treatment. Philadelphia: Lippincott. 158.

57. CONNOR W.E (1986) hyperlipidemic effects of diaterry w-3 fatty acids in normal and hyperlipidemic humans/ Effectiveness and mechanisms. In: SIMOPOULOS AP, KIFER RR, MARTIN R E, eds. Health effects of polyunsaturated fatty acids in seafoods. New York: Academic Press.173.

58. CODONER-FRANCH P, VALLS-BELLES P, BOIX L, TORRES MC, HERNANDEZ-MARCO R (2004) Is the obese child at risk of oxidative stress? SNFGE.P1.

59. CRUJEIRAS A B, PARRA M D, RODRIGUEZ M C, MARTINEZ DE MORENTIN B E, MARTINEZ J A (2006) A role for fruit content in energy-restricted diets in improving antioxidant status in obese women during weight loss. Nutrition. 22 (6): 593 - 599.

60. CUNNANE SC, GANGALI S, MENARD AC, LIEDE MJ, HAMEDEH ZY, CHEN TM., et al (1993) High α-linolenic acid flaxseed (Linum usitatissimum). Some nutritional properties in humans. British Journal of Nutrition. 69(2): 443-453.

61. CURTIN JF, DONOVAN M, COTTER TG (2002) Regulation and measurement of oxidative stress in apoptosis. J.Immunol. Methods. 265:49-72.

62. DARIMONT C, YURINI M, EPITAUX M, ZBINDEN I, RICHELLE M, MONTELL E, MARTINEZ AF, MACE K (2004) B3-adrenoceptor agonist prevents alterations of muscle diacylglycerol and adipose tissue phospholipids induced by a cafeteria diet. Nutri Metab. 1 :4-12.

63. DAUBRESSE J C, CADIÈRE G B, STERNON J (2005) Obesity in adult patients: check up and treatment. Rev Med Brux. 26: 33-42.

64. DAS U N, MOHAN I K, RAJU TR (2001) Effect of corticosteroids and eicosapentaenoic acid/docosahexaenoic acid on pro-oxidant and antioxidant statuts and metabolism of essential fatty acids in patients with glomerular disorder. Prostaglandins Leucot. Essent. Fatty.Acids. 65:197-203.

65. DAVID JS, BAZZAN A, WEAVER J, MAYER D, REICHLEF F.A (1987) cholesterol lowering mechanism by n-3 fatty acids in rat models. Arteriosclerosis. 7:535.

66. DE LORGERIL et al (1994) Mediterranean alphalinolenic acid-rich diet in secondary prevention of coronary heart disease. Lancet. 343(8951):1454.

67. DE SAINT POL T (2009) Evolution of obesity by social status in France, 1981-2003. Economics and human biology. 7 (3), 394-404.

68. DE ZWART LL, MEERMAN JHN, Commandeur J N, Vermeulen N P(1999) Biomarkers of free radical damage. Applications in experimental animals and in humans. Free Radic Biol Med .26: 202-26.

69. DEL VALLE LG (2011) Oxidative stress in aging: Theoretical outcomes and clinical evidences in humans. Biomedicine and aging pathology. 1:1-3

70. DELATTRE J, BEAUDEUX JL, BONNEFONT-ROUSSELOT D (2005) Radicaux libres et stress oxydant. Aspects biologiques et pathologiques. Cachan: Lavoisier.
71. DEMIGNÉ C, BLOCH-FAURE M, PICARD N, SABBOH H, BESSON C, RÉMÉSY C, GEOFFROY V, GASTON AT, NICOLETTI A, HAGÈGE A, MÉNARD J, MENETON P (2006). Mice chronically fed a westernized experimental diet as a model of obesity, metabolic syndrome and osteoporosis. Eur J Nutr. 45: 298-306.
72. DEMOZ A, WILLUMSEN N, BERGE R.K (1992) Eicosapentaenoid acid at hypotriglyceridemic dose enhances the hepatic antioxidant defense in mice. Lipids. 27:968-972.
73. DENKE MA (2001). Connections between obesity and dyslipidaemia. Curr Opin Lipidol. 12, 625-628.
74. DESPRES JP, LEMIEUX I (2006) Abdominal obesity and metabolic syndrome. Nature. 444:881-7.
75. DIDIER A, POSTIGO MA, MAILHOL C (2009) Asthma and obesity. Revue française d'allergologie. 49: 13-15.
76. DIETZE D, KOENEN M, RÖHRIG K, HORIKOSHI H, HAUNER H, ECKEL J (2002) Impairment of insulin signaling in human skeletal muscle cells by co-culture with human adipocytes. Diabetes. 51: 2369-2376.
77. DIETZE-SCHROEDER D, SELL H, UHLIG M, KOENEN M, ECKEL J (2005) Autocrine action of adiponectin on human fat cells prevents the release of insulin resistance-inducing factors. Diabetes. 54: 2003-2011.
78. DIXON JB (2010) The effect of obesity on health outcomes. Molecular and Cellular Endocrinology. 316(2):104-108.
79. DUPLUS E, GLORIAN M, TORDJMAN J, BERGE R, FOREST C (2002) Evidence for selective induction of phosphoenolpyruvate carboxykinase gene expression by unsaturated and nonmetabolized fatty acids in adipocytes. J Cell Biochem. 85:651-61.
80. DURAND G, GUESNET P, CHALON S, ALESSANDRI JM, RIZKALLA S, LEBRANCHU Y (2002) Importance nutritionnelle des acides gras polyinsaturés. In: Roberfroid M, Ed. Aliments Fonctionnels. Paris: Editions Tec & Doc-Lavoisier. 193-219.
81. DRAPER H, HADLEY M (1990) Methods Enzymol.186:421 -431.
82. DRENOWSKI A (1994). Human preference for sugar and fat. In: Fernstrom JD, Miller GD. Appetite and body weight regulation: sugar, fat and macronutrient substitutes. Boca Raton, Florida (USA), CRC Press. 137-147.
83. DREON DM, FERNSTROM HA, CAMPOS H, BL ANCHE P, WILLIAMS PT, KRAUSS RM (1998) Change in dietary saturated f a t intake is correlated with change in ma ss of large low-density-lipoprotein particles in me n. Am. J. Clin. Nutr. 67 :828-836.
84. ECKEL RH, YOST TJ, JENSEN DR (1995) Sustained weight reduction in moderately obese women results in decreased activity of skeletal muscle lipoprotein lipase. Eur J Clin Invest. 25: 396-402.
85. ELLMAN G L (1959) Tissue sulfhydryl groups. Arch. Biochem. Biophys. 82 :70-7.
86. ESTERBAUER H (1995) The chemistry of oxidation of lipoproteins. In: Rice-Evans C, Bruckdorfer KR (eds) oxidative stress. Lipoproteins and cardiovascular dysfunction.55- 79.
87. ESTERBAUER H, STREGL G, PUHL H, ROTHENEDER M (1989) In vitro oxidation of plasma lipoproteins. Continious monitoring of in vitro oxidation of human low density lipoprotein. Free. Radic. Biology. Medical. 6: 67 - 75
88. ESTEVE M, RAFECAS I, FERNANDEZ-LOPEZ JA, REMESAR X, ALEMANY M

(1994) Effect of a Cafeteria Diet on Energy Intake and Balance in Wistar Rats. Physiology & Behavior. 56: 65-71.

89. FAILOR R A, CHILDS M T, BIERMAN E (1988) The effects of n-3 and n-6 fatty acid enriched diets on plasma lipoprotein and apoproteins in familial combined hyperlipidemia. Metabolism. 37: 1021-1027.

90. FASHING P, RATHEISER , WALDHAUSL, W, ROHAC M, OSTERRODE W, NOWOTNEY P, VIERHAPPER H (1991) Metabolic effect of fish oil supplementation in patients with impared glucose tolerance. Diabetes. 40:583-589.

91. FERRANNINI E et al (1983) Effect of fatty acids on glucose production and utilization in man. J Clin Invest. 72 (5):1737-47.

92. FERREIRA LD, PULAWA LK, JENSEN DR, ECKEL RH (2001) Over expressing human lipoprotein lipase in mouse skeletal muscle is associated with insulin resistance. Diabetes. 50 : 1064-1068.

93. FERNANDEZ ML, SUN DM, MONTANO C, MCNAMARA DJ (1995) Carbohydrate-fat exchange and regulation of hepatic cholesterol and plasma lipoprotein metabolism in the guinea pig. Metabolism. 44: 855-864.

94. FERNANDEZ ML, VERGARA-JIMENEZ M, CONDE K, ABDEL-FATTAH G (1996) Dietary carbohydrate type and fat amount alter VLDL and LDL metabolism in guinea pigs. J. Nutr. 126: 2494-2504.

95. FICKOVA M, HUBERT P, CREMEL G, LERAY C (1998) Deatary (n-3) and (n-6) polyunsaturated fatty acids rapidly modify fatty acid composition and insulin effects in rat adipocytes.J.Nutr.128:512-519

96. FLACHS P, ROSSMEISL M, BRYHN M, KOPECKY J (2009*)* Cellularand molecular effects of n-3 polyunsaturated fatty acids on adipose tissue biology and metabolism. Clin Sci (Lond) 116:1-16.

97. FOLCH J, LEES M, SLOANE- STANLEY GH (1957) A simple method for isolation and purification of total lipids from animal tissues. J Biol Chem. 226 (1): 497-509.

98. FRANCIS DK, VANDEN BROECK J, YOUNGER N (2009). Fast food and sweetened beverage consumption: association with overweight and high waist circumference in adolescents. Public Health Nutr. 12 (8) :1106-1114.

99. FREEDMAN DS, SERDULA MK, SRINIVASAN SR, BERENSON GS (1999) Relation of circumferences and skinfold thicknesses to lipid and insulin concentrations in children and adolescents: the Bogalusa Heart Study. Am J Clin Nutr. 69(2):308-17.

100.FULOP T, LARBI A,WITKOWSKI JM, MCELHANEY J, LOEB M, MITNITSKIET A, PAWELEC G (2010) Aging, frailty and age-related diseases. Biogerontology. 11(5):547-563.

101.FUMERON F, BRIGHAN L , OLLIVER V, DE PROST D, DRISS F, DARCET P, BARD J M, PARRA HJ, FRUCHART J.C, APFELBAUM M (1991) n-3 polyunsatured fatty acids raise low-density lipoproteins, high density lipoprotein 2, and plasminogenactivator inhibitor in healthy young men. Am.J.Clin.Nutr.54:118-122.

102.FURUKAWA S, FUJITA T, SHIMABUKURO M, IWAKI M, YAMADA Y, NAKAJIMA Y,NAKAYAMA O, MAKISHIMA M, MATSUDA M, SHIMOMURA I (2004) Increased oxidative stress in obesity and its impact on metabolic syndrome. J. Clin. Invest. 114(12): 1752-1761.

103.GARCIA-DIAZ D, CAMPION J, MILAGRO F I, MARTINEZ J A (2007) Adiposity dependent apelin gene expression: relationships with oxidative and inflammation markers. Mol. Cell. Biochem. **27:** 432-444.

104.GASTALDELLI A, BALDI S, PETTITI M, TOSCHI E, CAMASTRA S, NATALI A, LANDAU BR, FERRANNINI E (2000) Influence of obesity and type 2 diabetes on gluconeogenesis and glucose output in humans: a quantitative study. Diabetes. 49(8):1367-73.
105.GAUVREAU D, VILLENEUVE N, DESHAIES Y, CIANFLONE K (2011) Novel adipokines: Links between obesity and atherosclerosis. Annales d'Endocrinologie.72 :224-231.
106.GELARDI N, LCHAC J, OH W (1990) Glucose metabolism in adipocytes of obese offspring of mild hyperglycemic clamp technique. Pediatr. Res.30: 40-44.
107.GESTA S, TSENG YH, KAHN CR (2007) Developmental origin of fat: tracking obesity to its source. Cell. 131:242-256.
108.GHISELLI A, SERAFINI M, NATELLA F, SCACCINI C (2000) total antioxidant capacity as a tool to assess redox status: critical view and experimental data. Free Radi. Biol.Med.29:1106-1114.
109.GIBSON RA, MUHLHAUSLER B, MAKRIDES M (2011) Conversion of linoleic acid and alpha-linolenic acid to long-chain polyunsaturated fatty acids (LCPUFAs), with a focus on pregnancy, lactation and the first 2 years of life. Maternal & Child Nutrition. 7(2):17-26.
110.GOLAY A, SWISLOCKI AL, CHEN YD, REAVEN GM (1987) Relationships between plasma-free fatty acid concentration, endogenous glucose production, and fasting hyperglycemia in normal and non-insulin-dependent diabetic individuals. Metabolism. 36(7):692-6.
111.GOLDBERG, 1996 GOLDBERG IJ (1996) Lipoprotein lipase and lipolysis: central roles in lipoprotein metabolism and atherogenesis. J Lipid Res. 37: 693-707.
112.GOTTO AMJ (1998) Triglyceride: The forgotten risk factor. Circulation. 97: 10271028.
113.GOURANTON E, LANDRIER JF (2007) Les facteurs modulant l'expression des adipokines en relation avec l'insulinorésistance associée à l'obésité. Obes. 2: 272-279.
114.GREGOIRE FM, SMAS CM, SUL HS (1998) Understanding adipocyte differentiation. Physiol Rev. 78: 783-809.
115.GRIMSRUD PA, PICKLO MJ, GRIFFIN TJ, BERNLOHR DA (2007). Carbonylation of adipose proteins in obesity and insulin resistance: identification of adipose fatty acidbinding protein as a cellular target of 4-hydroxynonenal. Mol Cell Proteomics. 6: 624637.
116.GROOP LC, SALORANTA C, SHANK M, BONADONNA RC, FERRANNINI E, DEFRONZO RA (1991) The role of free fatty acid metabolism in the pathogenesis of insulin resistance in obesity and noninsulin-dependent diabetes mellitus. J Clin Endocrinol Metab. 72(1):96-107.
117.GROUBET R, PALLET V, DELAGE B, REDONNET A, HIGUERET P, CASSAND P (2003) Hyperlipidic diets induce early alterations of the vitamin A signalling pathway in rat colonic mucosa. Endocrine regulations. 37: 137-144.
118.GRUNDY SM (2006) Diagnosis and management of the metabolic syndrome: an American Heart Association/National Heart, Lung, and Blood Institute scientific statement. Curr Opin Cardiol. 21 (1): 1-6.
119.HALLIWELL B, GUTTERIDGE JMC (1989) Free radicals in biology and medicine. 2nd ed. Oxford, UK: Clarendon.
120.HARIRI N, GOUGEON R, THIBAULT L (2010) A highly saturated fat-rich diet is more obesogenic than diets with lower saturated fat content. Nutrition Research. 30: 632-643.
121.HARMAN D (1956) Aging: a theory based on free radical and radiation chemistry. J Gerontol .11 :298-300.

122.HARRIS E.D (1992) Regulation of antioxidant enzymes. Fasebj. 6:2675-2683.
123.HARRIS WS (1989) Fish oils and plasma lipid and lipoprotein metabolism in humans: a critical review. J Lipid Res. 30: 785-807.
124.HARRIS WS (1990) omega-3 fatty acid effect on lipid metabolism. Cur.Op.Lipid. 1:5-11
125.HARRIS WS, CONNOR WE, ALAM N, ILLINGWORTH DR (1988) Reduction of postprandial triglyceridemia in humans by dietary n-3 fatty acids.J.lipid.Res.29:1451- 1457.
126.HAUG A, HOSTMARK A T (1987) Lipoprotein lipases, lipoproteins and tissue lipids in rats fed fish oil or coconut oil. J.Nutr.117:1011-1016.
127.HAUSMAN DB, DI GIROLAMO M, BARTNESS TJ, HAUSMAN GJ, MARTIN RJ (2001) The biology of white adipocyte proliferation. Obes. Rev. 2 : 239-254.
128.HIGDON JV, FREI B (2003) Obesity and oxidative stress: a direct link to CVD. Arteriosclr. Thromb. Vasc. Biol. 23: 365-367.
129.HIRSCH J, BATCHELOR B (1976) Adipose tissue cellularity in human obesity. Clin Endocrinol Metab 5: 299-311.
130.HOLGREM A (2003) Redox regulation of genes and cell function. In: Critical review of oxidative stress and aging. RG cutler and H Rodriguez Eds.World Scientific.2:102- 111.
131.HOLLIDAY R (2006) Aging is no longer an unsolved problem in biology. Ann NY Acad Sci. 1067: 1-9.
132.HOORROCKS LA, YEO YA (1999) Health benefits of docosahexaenoic acid (DHA). Pharmacol. Res. 40:211-225.
133.HOURIGAN R (2010) Cellular Energy Metabolism and Oxidative Stress. Text book of Aging Skin.Springer-Verlag Berlin Heidelberg. 30: 313-320.
134.HUSSEIN O, GOSOVSKI M, LASRI E, SVALB S, RAVID U, ASSY N (2007) World J Gastroenterol.13:361-368.
135.IRITANI N, KOMIYA M, FUKUDA H, SUGIMOTO T (1998) Lipogenic enzyme gene expression is quickly suppressed in rats by a small amount of exogenous polyunsaturated fatty acids. J Nutr. 128: 967-72.
136.JAESCHKE H (1995) Mechanism of oxidant stress-induced acute tissue injury. Proc. Soc. Exp. Biol.Med. 209:104-111.
137.JAMES WPT (2008) The epidemiology of obesity: the size of the problem. Journal of Internal Medicine. 263(4): 336-352.
138.JAYAKUMAR T, THOMAS PA, GERALDINE P (2007) Protective effect of an extract of the oyster mushroom, Pleurotus ostreatus, on antioxidants of major organs of aged rats. Exp. Gerontol. 42: 183-191.
139.JOCKEN W E, LANGIN D, SMIT E, SARIS W M, VALLE C, HUL G B, HOLM C, ARNER P, BLAAK E E (2007) Adipose triglyceride lipase and hormone-sensitive lipase protein expression is decreased in the obese insulin-resistant state. J. Clin. Endocrinol. Metab. **92**:2292-2299
140.JOLLY CA, JIANG YH, CHAPKIN R.S, MURAY DN (1997) dietary (n-3) polyunsatured fatty acids supress murine lymphoproliferation, Interleukin-2 secretion, and the formation of diacylglycerol and ceramide.J.Nutr.127:37-43.
141.JONES BH, MAHER MA, BANZ WJ, ZEMEL MB, WHELAN J, SMITH PJ, MOUSTAID N (1996) Adipose tissue stearoyl-CoA desaturase mRNA is increased by obesity and decreased by polyunsaturated fatty acids. AJP- Endo. 271(1):44-49.
142.JUCKER B M, CLINE G W, BARUCCI M, SHULMAN G I (1999) differential effect of sunflower oil versus fish oil feeding on insulin-stimulated glycogen synthesis, glycolysis, and

pyrovate dehydrogenase flux in skeletal muscle. Diabetes. 48:134-140.

143.JUNIEN C, GALLOU-KABANI C, VIGE A, GROSS MS (2005) Nutritional epigenomics of metabolic syndrome. Med/Sci. 21: 384-390.

144.KABBAJ O, YOON SR, HOLM C, ROSE J, VITALE ML, PELLETIER MR (2003) Biol Reprod. 68: 722-734.

145.KAITHWAS G, MAJUMDAR DK (2010) Therapeutic effect of *Linum usitatissimum* (flaxseed/linseed) fixed oil on acute and chronic arthritic models in albino rats. Inflammopharmacol 18:127-136.

146.KATHER H (1981) hormonal regulation of adipose tissue lipolysis in man: Implications for the pathogenesis of obesity.Triangle.20:131-143.

147.KELISHADI R, SHARIFI M, KHOSRAVI A, ADELI K (2007) Relationship between C-reactive protein and atherosclerotic risk factors and oxidative stress markers. Clin. Chem. 53: 456 - 464.

148.KELLE L, DIDIER G, LEANNE F (2009) Diet composition and obesity in adult Canadians. Health Reports. 20 : 1-4.

149.KELLEY DS, BRANCH LB, LOVE JF (1991) Dietary ALA and immunocompetence humans. American Journal of Clinical Nutrition. 53: 40-46.

150.KEMALI Z (2003) L'obésité au Maghreb. Santé Maghreb. December P1.

151.KESAVULU M, KAMESWARARAO B, APPARAO CH, KUMAR EG (2002) Effects of omega-3 fatty acids on lipid peroxidation and antioxidant enzyme status in type 2 diabetic and obese patients. Diabetes Metab. 28: 20-26.

152.KHAN NA, HICHAMI A (2002) Role of n-3 polyunsatured fatty acids in the modulation of T-cell signaling. In: Panadlai G (ed). Recent Research development in lipids. 6:65-78.

153.KIESS W, PITZOLD S, TOPFER M, GARTEN A, BLUHER S, KAPELLEN T, KORNER A (2008) Adipocytes and adipose tissue. Best Practice & Research Clinical Endocrinology & Metabolism. 22(1):135-153

154.KIM JK, GAVRILOVA O, CHEN Y, REITMAN ML, SHULMAN GI (2000) Mechanism of insulin resistance in A-ZIP/F-1 fatless mice. J Biol Chem. 275: 8456-60..

155.KISSEBAH AH, PEIRIS A N (1989) Biology of regional body fat distribution: relationship to non-insulin-dependent diabetes mellitus. Diabetes Metab Rev. 5(2):83- 109. Review.

156.KOPELMAN PG (2000) Obesity as a medical problem. Nature 404: 635-643.

157.KREGEL KC, ZHANG HJ (2007) An integrated view of oxidative stress in aging: basic mechanisms, functional effects, and pathological considerations. Am J Physiol Regul Integr Comp Physiol. 292: 18-36.

158.KUMARAN VS, ARULMATHI K, KALAISELVI P (2009) Senescence mediated redox imbalance in cardiac tissue: antioxidant rejuvenating potential of green tea extract. Nutrition, 25: 847-854.

159.LACROIX M, GAUDICHON C, MARTIN A, MORENS C, MATHE V, TOME D, HUNEAU JF (2004) A long-term high-protein diet markedly reduces adipose tissue without major side effects in Wistar male rats. Am J Physiol Regul Integr Comp Physiol. 287(4):934-42.

160.LANGIN D (2000) Millennium fat-cell lipolysis reveals unsuspected novel tracks. Horm. Metab. Res. 11:443-452.

161.LANGIN DL (2005) Adipocyte lipases and defect of lipolysis in human obesity.Diabetes. 11: 3190-3197.

162.LANE N (2003) Oxygen, the molecule that made the world. New York: Oxford University Press. 366 .
163.LARGE V (1998) Hormone-sensitive lipase expression and activity in relation to lipolysis in human fat cells. J Lipid Res. 25: 1688-1695.
164.LAVECCHIA C (2004) Mediterranean Diet and cancer. Public Health Nutr. 7:965968.
165.LE GOFF S, LEDEE N, BADER G (2008) Obesity and reproduction: A literature review. Obesity and reproduction: A literature review. Gynaecology Obstetrics & Fertility. 36:543-550.
166.LECERF JM (2008) Nutritional obesity. Obésité. 3(3):97-98.
167.LEE KW, LEE HJ and LEE CY (2002). Antioxidant activity of black tea vs. green tea. J Nutr .132 :785-786.
168.LEVINE RL, GARLAND D, OLIVER CN, AMICI A, CLIMENT I, LENZ AG, AHN BW, SHANTIEL S, STADTMAN ER (1990) Determination of carbonyl content in oxidatively modified proteins. Methods Enzymol.186:464-478.
169.LEVINE ET KIDD LEVINE RL, GARLEND D, OLIVIER CN, AMICI A, LENZ AG, SHANTIRL S, STADAN ER (1996) Determination of carbonyl content in oxidatively modified proteins. Methods Enzymol.186: 464-478.
170.LI Y, HOU M.J, MA J, TANG Z H, ZHU HL, LING WH (2005) dietary fatty acids regulate cholesterol induction of liver CYP7alphal expression and bile acid production. Lipids (50):455-462.
171.LIADO I, PROENZA AM, SERRA F, PALOU A, PONS A (1991) Dietary-induced permanent changes in brown and white adipose tissue composition in rats. Int J Obesity. 15: 415-419.
172.LICHTENSTEIN AH, KENNEDY E, BARRIER P (1998) Dietary fatty consumption and health.Nutr Rev. 56: 3-19.
173.LISSNER L, HEITMANN BL (1995) Dietary fat and obesity: evidence from epidemilogy. European journal of Clinical Nutrition. 49 : 79-90.
174.LONDOÑO-VALLEJO JA (2009) A Nobel centenarian celebrates telomeres and telomerase. Med Sci (Paris).25:973-6
175.LOPEZ IP, MARTI A, MILAGRO FI (2003) DNA microarray analysis of genes differentially expressed in diet induced (cafeteria) obese rats. Obes Res.11:188-194.
176.LORENTE-CEBRIÁN S, COSTA AG, NAVAS-CARRETERO S, ZABALA M, MARTÍNEZ JA, MORENO-ALIAGA MJ (2013) Role of omega-3 fatty acids in obesity, metabolic syndrome, and cardiovascular diseases: a review of the evidence. J Physiol Biochem. 69(3):633-51
177.LOUGHEED M, STEINBRECHER UP (1996) Mechanism of uptake of copper- oxidized low density lipoprotein in macrophages is dependent on its extent of oxidation. J Biol Chem. 271(20): 805-11798.
178.LOUIS-SYLVESTRE J (1984) Mécanisme de l'induction de l'hyperphagie et de l'obésité par le regime cafeteria. Cahier de nutrition et de diététique. 4: 197-204.
179.LOVEJOY JC (2002) "The influence of dietary fat on insulin resistance." Curr Diab Rep 2(5): 400-435.
180.LOWRY OH, ROSENBROUGH NJ, FARR AL, RANDALI RI (1951) Protein measurement with folin phenol reagent. J Biol Chem. 193: 265-275.
181.MAGKOS F, MOHAMMED BS, MITTENDORFER B (2009) Plasma Lipid Transfer Enzymes in Non-Diabetic Lean and Obese Men and Women. Lipids. 44(5):459-464.

182.MAIESE K, CHONG Z Z., et al (2007) "Mechanistic insights into diabetes mellitus and oxidative stress." Curr Med Chem 14(16): 1729-38.

183.MAKNI M, FETOUI H, GARGOURI N K, GAROUI EL M, JABER H, MAKNI J, BOUDAWARA T, ZEGHAL N (2008) Hypolipidemic and hepatoprotective effects of flax and pumpkin seed mixture rich in n-3 and n-6 fatty acids in hypercholesterolemic rats. Food and Chemical Toxicology 46:3714-3720

184.MASEK J, FABRY P (1959) High-fat diet and the development of obesity in albino rats.Experientia. 15: 444-445.

185.MATHE D, SEROUGNE C, FEREZOU J, LECUYER B (1991) Ann Nutr Metab 35, 165-173.

186.MAZZA G, OOMMAH B D (2000) Functional foods, Biochemical and Processing Aspects. Technomic. Publ. Co Inc, Lancaster.

187.MAZZUCOTELLI A, LANGIN D (2006) La mobilisation des acides gras et leur utilisation dans le tissu adipeux : une nouvelle donne. Journal de la Société de Biologie. 200 (1) 83-91.

188.MCGARRY JD (2002) dysregulation of fatty acid metabolismin the etiology of type 2 diabetes. Diabetes.51:7-18.

189.MERZOUK H, KHAN N A (2003) Implication of lipids in macrosomia of diabetic pregnancy: Can n-3 Polyinsatured fatty acids exert beneficial effect. Clin.Sci.105:519- 529.

190.MERZOUK H, MADANI S, BOUALGA A, PROST J, BOUCHENAK M, BELEVILLE J (2001) Age-related changes in cholesterol metabolism in macrosomic offspring of rats with streptozotocin-induced diabetes. J Lipid Res.42:1152-1159.

191.MERZOUK S, HICHAMI A, MADANI S, MERZOUK H, YAHIA BERROUIGUET A, PROST J, MONTAIRO K, CHABANE SARI N, KHAN NA (2003) Antioxidant status levels of different vitamins determined by high performance liquid chromatography in diabetic subjects with complications. Gen Physiol Biophys. 22: 1557.

192.MERZOUK S, HICHAMI A, SARI S, MADANI S, MERZOUK H, YAHIA BERROUIGUET A,LENOIR-ROUSSEAUX J, CHABANE SARI N, KHAN NA (2004) Impaired oxidant/Antioxidant status and LDL-Fatty Acid Composition are associated with increased susceptibility to peroxidation of LDL in diabetic patients.Gen. Physiol. Biophys. 23: 387-399.

193.MICHALIK, DESVERGNE B, WAHLI W (2000) Les bases moléculaires de l'obésité : vers de nouvelles cibles thérapeutiques. Med/Sci. 16: 1030-1039.

194.MICHEL F, BONNEFONT-ROUSSELOT D, MAS E, DRAI J, THEROND P (2008) Biomarkers of lipid peroxidation: analytical approaches. Ann Biol clin. 66(6): 605-20.

195.MILARGO FI, CAMPION J, MARTINEZ JA (2006) Weight gain induced by high fat feeding involves increased liver oxidative stress. Obesity. 14: 1118-1123.

196.MILES E.A, CALDER P.C (1998) Modulation of immune function by dietary fatty acids.Proc.Nut.Soc.57:277-292.

197.MIRET S, SAIZ MP, MITJAVILA MT (2003) Br J Nutr 89 : 11-18.

198.MIYOSHI H, SOUZA SC, ZHANG HH, STRISSEL KJ, CHRISTOFFOLETTE MA, KOVSAN J, RUDISH A (2006) Perilipin promotes hormone-sensitive lipase mediated adipocyte lipolysis via phosphorylation dependent and independent mechanisms. J Biol Chem. 281 : 15837-44.

199.MOORE K, ROBERTS II LJ (1998) Measurement of lipid peroxidation. *Free Rad Res.*28 : 659-71

200.NESTEL PJ, WONG S, TOPPING DL (1986) Dietary long chain polyenoic fatty acids: Suppression of triglyceride formation in rat liver. Acad.press.211-218.
201.NESTELET PJ, CONNOR WE, REARDON ME, CONNOR S, WONG S, BOSTON R (1984) Suppression by diets rich in fish oil of very low density lipoprotein production in man.J.Clin.Invest.74:82-89.
202.NORDOY et al (2001) n-3 polyunsaturated fatty acids and cardiovascular diseases. Lipids. 36: 127-9.
203.NOUROOZ-ZADEH J, TAJADDINI-SARMADI J, LINKLE, WOLFF SP (1996) Low density lipoprotein is the major carrier of lipid hydroperoxides in plasma. Biochem J. 313: 781-786.
204.OAKES N.D, COONEY G.I, CAMILLERI S (1997) Mechanisms of liver and muscle insulin resistance induced by chronic high-fat feeding. Diabetes.46: 1768-1774.
205.OLUSI SO (2002) obesity is an independent risk factor for plasma lipid peroxidation and depletion of erythrocyte cytoprotectic enzymes in human. Int J Obes relat metab Disord.26 (29):1159-1164.
206.ONG JM, KERN PA (1989) Effect of feeding and obesity on lipoprotein lipase activity, immunoreactive protein, and messenger RNA levels in human adipose tissue. J Clin Invest. 84: 305-311.
207.OOMAH BD (2001) Flaxseed as a functional food source. Journal of the Science of Food and Agriculture. 81: 889-894.
208.OWUOR ED , KONG AN (2002) antioxidants and oxidants regulated signal transduction pathways. Biochem. Pharmacol. 64:756-770.
209.PAN A, YU D, DEMARK-WAHNEFRIED W, FRANCO OH, LIN X (2009) Metaanalysis of the effects of flaxseed interventions on blood lipids. Am J Clin Nutr. 90: 28897.
210.PASQUET P, FRELUT ML, SIMMEN B, HLADIK CM, MONNEUSE MO (2007) Taste perception in massively obese and in non-obese adolescents. Int J Pediatr Obes. 2(4):242-8.
211.PENICAUD L, COUSIN B, LELOUP C, LORSIGNOL A, CASTEILLA L (2000) The autonomic nervous system, adipose tissue plasticity and energy balance, Nutrition. 16 : 903-908.
212.PERRIN D (2003) Characterisation of energy balance and its central and peripheral monaminergic control in a rat model that does not develop obesity, the LOU/C rat. Doctoral thesis, Université CLAUDE BERNARD LYON1.
213.PERTICONE F, CERAVOLO R, CANDIGLIOTA M, VENTURA G, IACOPINO S, SINOPOLI F, MATTIOLI PL (2001) Obesity and body fat distribution induce endothelial dysfunction by oxidative stress: Protective effect of vitamin C. Diabetes. 50 (1):159-165.
214.PETIT V, ARNOULD L, MARTIN P, MONNOT MC, PINEAU T, BESNARD P, NIOT I (2007) Chronic high-fat diet affects intestinal fat absorption and postprandial triglyceride levels in the rate. J Lipid Res. 48: 278-287.
215.POUBELLE P, CHAINTREUIL J, BENSADOUN J, BLOTMAN F, SIMON L, CRASTES DE PAULET A (1982) Plasma lipoperoxides and aging. *Biomedicine* .36 : 164-7.
216.POUDYAL H, PANCHAL SK, DIWAN V, BROWN L (2011) Omega-3 fatty acids and metabolic syndrome: effects and emerging mechanisms of action. Prog Lipid Res 50:372-87.
217.POUDYAL H, PANCHAL SK, WARD LC, BROWN L (2013) Effects of ALA, EPA and DHA in high-carbohydrate, high-fat diet-induced metabolic syndrome in rats. J Nutr

Biochem. 24(6):1041-52.

218.POULIOT MC et al (1992) Visceral obesity in men. Associations with glucose tolerance, plasma insulin, and lipoprotein levels. Diabetes 41, 826-834 .

219.RAJASEKARAN NS, DEVARAJ H, DEVARAJ SN (2002) The effect of glutathione monoester (GME) on glutathione (GSH) depleted rat liver. J. Nutr. Biochem. 13: 302306.

220.RATTAN SI (2008) Increased molecular damage and heterogeneity as the basis of aging. Biol Chem. 389: 267-72.

221.REAVEN G (2005) All obese individuals are not created equal: insulin resistance is the major determinant of cardiovascular disease in overweight/obese individuals. Diab Vasc Dis Res **2**(3): 105-12.

222.ROBERT H, WANG H (2009) Lipoproteine lipase: from gene to obesity. Am J Physiol Endocrinol Metab. 10 : 1152.

223.ROBINSON LE, BUCHHOLZ AC, MAZURAK VC (2007) Inflammation, obesity, and fatty acid metabolism: influence of n-3 polyunsaturated fatty acids on factors contributing to metabolic syndrome. Appl Physiol Nutr. Metab. 32:1008-24.

224.ROE JH, KUETHER CA (1943) The determination of ascorbic acid in whole blood and urine through the 2, 4- dinitrophenylhydrazine derivatives of dehydroascorbic acid. J Biol Chem.14:399-407.

225.ROMANO A D, SERVIDDIO G, et al (2010) "Oxidative stress and aging." J Nephrol 23 Suppl 15: S29-36.

226.ROMERO AL, FERNANDEZ ML (1996) Dietary fat amount and carbohydrate type regulate hepatic acyl CoA: cholesterol acyltransferase (ACAT) activity. Possible links between ACAT activity and plasma cholesterol levels. Nutr. Res. 16 : 937-948.

227.ROSEN ED, MACDOUGALD OA (2006) Adipocyte differentiation from the inside out. Nat Rev Mol Cell Biol .7: 885-896.

228.RUSTAN AC, HUSTVEDT BE, DREVON CA (1993) Dietary supplementation of very long-chain n-3 fatty acids decreases whole lipid utilization in the rat. J Lipid Res. 34: 1299-1309.

229.RUTH MR, TAYLOR CG, ZAHRADKA P, FIELD CJ (2008). Abnormal Immune Responses in fa/fa Zucker Rats and Effects of Feeding Conjugated Linoleic Acid. Obesity. 16: 1770-1779.

230.SACKS FM, KATAN M (2002) Randomized clinical trials on the effects of dietary fat and carbohydrate on plasma lipoproteins and cardiovascular disease. Am. J. Med. 113: 13S-24S.

231.SADUR CN, YOST TJ, ECKEL RH (1984) Insulin responsiveness of adipose tissue lipoprotein lipase is delayed but preserved in obesity. J Clin Endocrinol Metab. 59: 1176-1182.

232.SAELY CH, GEIGER K, DREXEL H (2012) Brown versus white adipose tissue: a mini-review. Gerontology 58(1):15-23.

233.SAKURAI Y (2000) Duration of obesity and risk of non-insulin-dependent diabetes mellitus. Biomed. Pharmacother 54, 80-84.

234.SALTER AM, MANGIAPANE EH, BENNETT AJ, BRUCE JS, BILLETT MA, ANDERTON KL, MARENAH CB, LAWSON N, WHITE DA (1998) The effect of different dietary fatty acids on lipoprotein metabolism: concentration- dependent effects of diets enriched in oleic, myristic, palmitic and stearic acids. Br. J. Nutr. 79: 195-202.

235.SANDERS TA, ROSHANAI F (1983) The influence of different types of w-3 polyunsaturated fatty acids on blood lipids and platelet function in healthy volunteers.Clin.Sci.64:91-96.

236.SARSILMAZ M, SONGUR A, OZYURT H, KUS I, OZENO A, OZYURT B (2003). Potential role of dietary omega 3 essential fatty acids on some oxidant/antioxidant parameters in rats' corpus striatum. Prostaglandins leukot. Essent Fatty Acids. 69:253259.

237.SCHAEFER EJ, LEVY RI (1985) Pathogenesis and management of lipoprotein disorders. N Engl J Med.312: 1300-1310.

238.SCHLIENGER JL, LUCA F, PRADIGNAC A (2010) What's new in the treatment of obesity? What's new in the treatment of obesity? La Revue de médecine interne. 31:185-187.

239.SCHAEFER EJ, LEVY RI (1985) Pathogenesis and management of lipoprotein disorders. N Engl J Med. 312, 1300-1310.

240.SCHWARTZ RS, BRUNZELL JD (1981) Increase of adipose tissue lipoprotein lipase activity with weight loss. J Clin Invest. 67: 1425-1430.

241.SCOARIS CR, VASCONCELOS RIZO G, ROLDI LP, FRANZOI DE MORAES SM, GOMES DE PROENC AR, PERALTA RM, MARÇAL NMR (2010). Effects of cafeteria diet on the jejunum in sedentary and physically trained rats. Nutrition. 26: 312320.

242.SENTHIL KUMARAN V, ARULMATHI K, SRIVIDHYA R, KALAISELVI P (2008) Repletion of antioxidant status by EGCG and retardation of oxidative damage induced macromolecular anomalies in aged rats. Exp. Gerontol. 43: 176-183.

243.SHAFAT A, MURRAY B, RUMSEY D (2009) Energy density in cafeteria diet induced hyperphagia in the rat. Appetite. 52: 34-38.

244.SHERHAN CN, ARITA M, HONG S, GOTLING K (2004) Resolvins, Docosatrienes and neuroprotectins, novel àmega-3 derived mediators and their endogenous aspirin- tiggered epimers.Lipids.39:1125-1132.

245.SHIROUCHI B, NAGAO K, INOUE N, OHKUBO T, HIBINO H, YANAGITA T (2007) Effect of dietary omega 3 phosphatidylcholine on obesity-related disorders in obese Otsuka Long-Evans Tokushima fatty rats. J Agric Food Chem. 55:71-76.

246.SHULMAN GI (2000) Cellular mechanisms of insulin resistance. J Clin Invest.106: 171-6.

247.SIES H (1997) Antioxidant in disease mechanisms and therapy, Advences in Pharmacology. Academic. Press. New York. 38.

248.SIMOPOULOS AP, ROBINSON J (1998) The Omega plan. New York, Harper Collins. 1-55.

249.SINGER P, JAEGER W, BERGER, BARLEBEN H, WIRTH M, RICHTERHEIRICH E, VOIGT S, GODICKE W (1990) Effects of dietary oleic, linoleic and alpha-linolenic acids on blood pressure, serum lipids, lipoproteins and the formation of eicosanoid precursors in patients with mild essential hypertension.J.Hum.Hypertens.4(3):227-233.

250.SINGH RB, BEEGOM R , RASTOGI SS, GAOLI Z, SHOUMIN Z (1998) Association of low plasma concentrations of antioxidant vitamins, magnesium and zinc with high body fat percent measured by bioelectrical impedance analysis in Indian men. Magnes Res. 11 (1): 3 - 10.

251.SINGH RB, NIAZ M A, BISHNOI I, SHARMA JP, GUPTA S, RASTOGI SS, SINGH R, BEEGOM R, CHIBO H, SHOUMIN Z (1994) Diet, antioxidant vitamins, oxidative stress and risk of coronary artery disease: the Peerzada Prospective Study. *Acta Cardiol.* 49 (5): 453 - 467.

252.SIRIWARDHANA N et al (2013) J. Nutr. Biochem. 24 : 613-623

253.SIRTORI CR. AND C. GALLI (2002) "N-3 fatty acids and diabetes." Biomed Pharmacother. 56(8): 397-406.
254.SLOVER H T, LANZA E (1979) Quantitative analysis of food fatty acids by cappilary gas chromatography. J Am Oil Chem Soc. 56: 933-943.
255.SOULIMANE-MOKHTARI N, GUERMOUCHE B, YESSOUFOU A, SAKER M, MOUTAIROU K, HICHAMI A, MERZOUK H, KHAN NA (2005) Modulation of lipid metabolism by (N-3) PUFA in gestational diabetic rats and their macrosomic offspring. Clin Sci. 109 : 287-295.
256.SREEKUMAR R, UNNIKRISHNAN J, FU A, NYGREN J, SHORT KR, SCHIMKE J et al (2002) Impact of high-fat diet and antioxidant supplement on mitochondrial functions and gene transcripts in rat muscle. Am. J. Physiol Endocrinol Metab. Metab. 282: 1055-1061.
257.STORLIEN LH, KRAEGEN E W, GHISHOLM DJ, FORD GL, BRUCE DG, PASCOE WS (1987) Fish oil prevents insulin resistance induced by high-fat feeding in rats.Science.237:885-888.
258.STORLIEN LH, HULBERT AJ, et al (1998) "Polyunsaturated fatty acids, membrane function and metabolic diseases such as diabetes and obesity." Curr Opin Clin Nutr Metab Care. 1(6): 559-63
259.STUMVOLL M, MEYER C, MITRAKOU A (1997) Renal glucose production and utilization: new aspects in humans. Diabetologia. 40:749-57.
260.STURM R (2007) Increases in morbid obesity in the USA: 2000-2005. Public Health, 121(7): 492-496.
261.SUBBAIAH P.V, SUBRAMANIAN VS, LIU M (1998) Trans unsaturated fatty acids inhibit lecithin: cholesterol acyltransferase and alter its positional specificity. J. Lipid. Res. 39 (7):1438-1447.
262.SURESH Y, AND U, DAS N (2003) "Long-chain polyunsaturated fatty acids and chemically induced diabetes mellitus: effect of omega-6 fatty acids." Nutrition .19(2): 93-114.
263.SZOKE E, SHRAYYEF MZ, MESSING S, WOERLE HJ, VAN HAEFTEN TW, MEYER C, MITRAKOU A, PIMENTA W, GERICH JE (2007) Effect of aging on glucose homeostasis accelerated deterioration of β-cell function in individuals with impaired glucose tolerance. Diabetes Care. 31(3):539-543.
264.TAYLOR F (1985) Flow-trought PH-stat method for lipase activity. Analytical biochemistry.148:149-153.
265.TEIXEIRA PJ, SARDINHA LB, GOING SB, LOHMAN TG (2001) Total and regional fat and serum cardiovascular disease risk factors in lean and obese children and adolescents. Obes Res. 9(8):432-42.
266.TIETZ NW, ASTLES JR, SHUEY DF (1989) Lipase activity measured in serum by a continuous-monitoring pH-stat technique -an update. Clin chem.35:1688-1693.
267.TILG H, MOSCHEN AR. (2006) Adipocytokines: mediators linking adipose tissue, inflammation and immunity. Nat Rev Immunol. 6: 772-783.
268.TSUKAMOTO Y, OKUBO M, YONEDA T, MARUMO F, NAKAMURA H (1982) Effects of a polyunsaturated fatty acid-rich diet on serum lipids in patients with chronic renal failure.Nephron.31(3):236-241.
269.TZANG BS, YANG SF, FU SG, YANG HC, SUN HL, CHEN YC (2009). Effects of dietary flaxseed oil on cholesterol metabolism of hamsters. Food Chemistry. 114: 14501455.
270.UNGER RH (2003) Minireview: weapons of lean body mass destruction: the role of ectopic lipids in the metabolic syndrome. Endocrinology. 144(12):5159-65.

271.UZUN H, KONUKOGLU D, GELISGEN R, ZENGIN K, TASKIN M (2007) Plasma protein carbonyl and thiol stress before and after laparoscopic gastric banding in morbidly obese patients.Obes Surg. 17: 1367-1373.
272.VENKATRAMAN JT, CHANDRASEKAR B, KIM JD, FRNADES G (1994) Effect of n-3 and n-6 fatty acids on the activities and expression on hepatic antioxidant enzymes in autoimmune-prone NZBxNZWFI mice.Lipids.29:561-568.
273.VERGES B (2001) Insulinosensitivity and lipids. Diabetes Metab. 27 : 223-227.
274.VIDON C, BOUCHER P, CACHEFO A, PERONI O, DIRAISON F, BEYLOT M (2001) Effects of isoenergetic high-carbohydrate compared with high-fat diets on human cholesterol synthesis and expression of key regulatory genes of cholesterol metabolism. Am. J. Clin. Nutr. 73: 878-884.
275.VIJAIMOHAN K, JAINU M, SABITHA KE, SUBRAMANIYAM S, ANANDHAN C, DEVI CSS (2006) Beneficial effects of alpha linolenic acid rich flaxseed oil on growth performance and hepatic cholesterol metabolism in high fat diet fed rats. Life Sciences. 79: 448-54.
276.VINCENT HK, INNES K E, et al (2007) "Oxidative stress and potential interventions to reduce oxidative stress in overweight and obesity. "Diabetes Obes Metab. 9(6): 81339.
277.VINCENT, HK, TAYLOR AG (2006) Biomarkers and potential mechanisms of obesityinduced oxidant stress in humans. *Int. J. Obes* (Lond) 30:400-418,.
278.VINER RM, SEGAL TY, LICHTAROWICZ KE, HINDMARSH P (2005) Prevalence of the insulin resistance syndrome in obesity *.Arch. Dis. Child.* 90: 10-14.
279.WAKIL S J (1989) The fatty acid synthase: A proficient multifunctional enzyme. *Biochemistry.* **28:** 4523-4530.
280.WAKIL SJ, ABU-ELHEIGA LA (2009) Fatty acid metabolism: target for metabolic syndrome. *J. lipid. Res.* 1194: 200-215.
281.WANDER RC, SHI-HUA D U (2000) Oxidation of plasma proteins is not increased after supplementation with eicosapentaenoic acids. Am. J.Clin.Nutr. 72 :731-737.
282.WESLY A (1998) Immunonutrition: The role of n-3 fatty acid. Science (Washington D.C). 14:627-633.
283.WEST DB, YORK B (1998) Dietary fat, genetic predisposition, and obesity: Insights from animal models. Am J Clin Nutr. 67: 505-512.
284.WHITLOCK G, LEWINGTOW S, SHERLIKER P, CLARKE R, EMBERSON J, HALSEY J (2009) Body-mass index and cause- specific mortality in 900000 adults: collaborative analyses of 57 prospective studies. Lancet. 373: 1083-96.
285.WIERNSPERGER NF (2003) "Oxidative stress as a therapeutic target in diabetes: revisiting the controversy." Diabetes Metab. 29(6): 579-85.
286.WONG SH, MARSH SB (1988) Differential effects of eicosapentaenoic and docosahexaenoic acids on triacylglycerol and apolipoprotein B production in HEP G2 Cells. Atherosclerosis. 8:63.
287.WONG S, REARDON M, NESTEL P (1985) Reduced triglyceride formation from long-chain polyenoic fatty acids in rat hepatocytes. Metabolism.34:900-905.
288.WONG, S,H, MARSH JB,(1988) Differential effect of eicosapentaenoiec and docosahexaenoiec acids on triacylglycerol and apolipoprotein B production in HEP G2 cells. Atherosclerosis.8:63.
289.WORLD HEALTH ORGANISATION (2010) Set of recommendations on the marketing of food and non-alcoholic beverages to children. Geneva: WHO.

http://whqlibdoc.who.int/publications/2010/9789242500219_fre.pdf

290.WORLD HEALTH ORGANIZATION (1998) Obesity: preventing and managing the global epidemic, in: WHO, Report of a WHO Consultation on Obesity (WHO/NUT/NCD/98.1), Geneva, Switzerland.

291.WORLD HEALTH ORGANISATION (2000) Obesity: Preventing and Managing the Global Epidemic. WHO Obesity Technical Report Series, 894.

292.XIN W, WEI W, LI XY(2013) Short-term effects of fish-oil supplementation on heart rate variability in humans: a meta-analysis of randomized controlled trials. Am J Clin Nutr. 97(5):926-35.

293.YAZU K, YAMAMOTO Y, NIKI E, MIKI K, UKEGAWA K (1998) Mechanisms of lower oxidizability of eicosapentaenoate than linoleate in aqueous micelles. Il. Effect of antioxidants.Lipids.33:597-600.

294.YESSOUFOU A, SOULIMANE N, MERZOUK SA, MOUTAIROU K, AHISSOU H, PROST J, SIMONIN A.M, MERZOUK H, HICHAMI A, KHAN N.A (2006) n-3 fatty acids modulate antioxidant status in rats and their macrosomic offspring. International Journal of obesity.1-12

295.YILMAZ O, OZKAN Y, YILDIRIM M, OZTURK AI, ERSAN Y (2002) Effects of alpha lipoic acid, ascorbic acid-6-palmitate, and fish oil on glutathione, malonaldehyde and fatty acid levels in erythrocytes of diabetic male rats.J.Cell.Biochem.86:530-539

296.YOUSSEF H, GROUSSARD C, PINCEMAIL J, MOUSSA E, ZIND M, LEMOINE S, JEAN-OLIVIER D, CILLARD J, DELAMARCHE P, GRATAS-DELAMARCHE A (2009) Post-exercise oxidative stress in overweight adolescent girls: implication of basal insulin-resistance and inflammation exacerbate. International Journal of Obesity. 33: 447-455.

297.YU C, CHEN Y, CLINE GW, ZHANG D, ZONG H, WANG Y, BERGERON R, KIM JK, CUSHMAN SW, COONEY GJ, ATCHESON B, WHITE MF, KRAEGEN EW, SHULMAN GI (2002) Mechanism by which fatty acids inhibit insulin activation of IRS-1 associated phosphatidyl inositol 3-kinase activity in muscle. J Biol Chem. 277:50230-50236.

298.YUAN YV, KITTS D.D (2002) dietary fat source and cholesterol interactions alter plasma lipids and tissue susceptibility to oxidation in spontaneously hypertensive (SHR) and normotensive Wistar Kyoto(wky) rats .Mol.Cell.Biochem.232:33-47

299.ZARROUKI B, SOARES A F et al (2007) "The lipid peroxidation end-product 4- HNE induces COX-2 expression through p38MAPK activation in 3T3-L1 adipose cell." FEBS Lett. 581(13): 2394-400.

APPENDICES

Lots	Control rats			Obese rats			P (ANOVA)
Parameters	C	CL2.5% (%)	CL5%	CAF	CAFL2.5	CAF5%.	
Body weight (g)	521.66±5.13[c]	491.43±4.36[d]	443.33±4.27[e]	600.25±5.23[a]	552.71±7.31[b]	515.30±6.22[c]	0,0001
Food consumption (g/d/rat)	39.08±1.42[b]	22.53±1.30[c]	21.88±2.54[c]	45.18±2.37[a]	21.48±2.32[c]	21.95±1.77c	0,001
Energy intake (Kcal/d/rat)	150,21±6.42[b]	138.33±7.30[c]	122.43±3.30[c]	230.58±7.30[a]	152.43±4.30[c]	151.04±3.30[b]	0,01

Each value represents the mean ± SD, n=10.C: control rats fed the standard diet; CAF: obese rats fed the cafeteria diet; CL2.5%: control rats fed the standard diet enriched with 2.5% linseed oil; CAFL2.5%: obese rats fed the cafeteria diet enriched with 2.5% linseed oil; CL5%: control rats fed the standard diet enriched with 5% linseed oil; CAFL5%: obese rats fed the cafeteria diet enriched with 5% linseed oil. After verifying the normal distribution of the variables (Shapiro-Wilk test), the means of the six groups of rats were compared using a one-factor ANOVA test. This analysis was completed by the Tukey test in order to classify and compare the means in pairs. Means indicated by different letters (a, b, c) are significantly different (p<0.05).

Lots	Control rats			Obese rats			P (ANOVA)
Parameters	C	CL2.5% FOR	CL5%	CAF	CAFL2.5	CAF5%.	
Glucose (g/l)	1,13±0,15 [d]	1,03±0,07 [d]	1±0,12 [d]	1,90 ±0,10 [a]	1,60±0,09 [c]	1,33 ±0,22[c]	0,01
Urea (g/l)	0,27±0,06 [b]	0,26±0,03 [b]	0,24±0,04 [b]	0,54±0,09 [a]	0,24±0,02 [b]	0,27±0,02 [b]	0,01
Creatinine (g/l)	13,45±1,77[b]	13,49±2,10[b]	13,56±2,48[b]	21,75±4,17[a]	15,59±3,76[b]	14,79±2,03[b]	0,01

Each value represents the mean ± SE, n=10.C: control rats fed the standard diet; CAF: obese rats fed the cafeteria diet; CL2.5%: control rats fed the standard diet enriched with 2.5% linseed oil; CAFL2.5%: obese rats fed the cafeteria diet enriched with 2.5% linseed oil; CL5%: control rats fed the standard diet enriched with 5% linseed oil; CAFL5%: obese rats fed the cafeteria diet enriched with 5% linseed oil. After verifying the normal distribution of the variables (Shapiro-Wilk test), the means of the six groups of rats were compared using a one-factor ANOVA test. This analysis was completed by the Tukey test in order to classify and compare the means in pairs. Means indicated by different letters (a, b, c) are significantly different (p<0.05).

Lots	Control rats			Obese rats			P ANOVA
Parameters	C	CL2.5% FOR	CL5%	CAF	CAFL2.5	CAF5%.	
Serum (mg/dl)	169,03±0,06 [b]	162,3±0,05 [b]	155,2±0,02 [b]	200,21±0,12 [a]	174,13 ±0,05 [b]	166,12±0,04[b]	0,01
VLDL (mg/dl)	32,15 ±0,03 [b]	27,2±0,01[c]	25,3±0,03 [c]	49,22±0,02 [a]	43,13±0,01 [c]	27,45±0,01 [c]	0,01
LDL (mg/dl)	51,4±0,02 [b]	37,56±0,03[c]	25,23±0,03 [d]	64,09±0,02 [a]	42,12±0,05 [c]	40,23±0,01 [c]	0,01
HDL (mg/dl)	45,34±0,03[b]	57,02±0,02[a]	61,47±0,02[a]	28,90±0,04[c]	46,23±0,01[b]	50±0,02[b]	0,01

Each value represents the mean ± SE, n=10.C: control rats fed the standard diet; CAF: obese rats fed the cafeteria diet; CL2.5%: control rats fed the standard diet enriched with 2.5% linseed oil; CAFL2.5%: obese rats fed the cafeteria diet enriched with 2.5% linseed oil; CL5%: control rats fed the standard diet enriched with 5% linseed oil; CAFL5%: obese rats fed the cafeteria diet enriched with 5% linseed oil. After verifying the normal distribution of the variables (Shapiro-Wilk test), the means of the six groups of rats were compared using a one-factor ANOVA test. This analysis was completed by the Tukey test in order to classify and compare the means in pairs. Means indicated by different letters (a, b, c) are significantly different (p<0.05).

Lots Parameters	Control rats			Obese rats			P (ANOVA)
	C	CL2.5% FOR	CL5%	CAF	CAFL2.5	CAFL5%	
Serum (mg/dl)	138±0,04[b]	134 ±0,03 [c]	125±0,04[d]	173±0,14 [a]	142±0,02 [b]	136±0,02 [c]	0,005
VLDL (mg/dl)	54±0,02 [b]	41±0,03 [c]	37±0,04 [d]	67±0,04 [a]	45±0,02 [c]	41±0,01 [c]	0,001
LDL (mg/dl)	24±0,05 [b]	22±0,05[b]	20±0,04[c]	27±0,05[a]	24±0,04[b]	23±0,03[b]	0,01
HDL (mg/dl)	37 ±0,02[b]	26 ±0,01[c]	25±0,04[d]	44±0,03[a]	34±0,02[b]	31±0,01[b]	0,01

Each value represents the mean ± SE, n=10.C: control rats fed the standard diet; CAF: obese rats fed the cafeteria diet; CL2.5%: control rats fed the standard diet enriched with 2.5% linseed oil; CAFL2.5%: obese rats fed the cafeteria diet enriched with 2.5% linseed oil; CL5%: control rats fed the standard diet enriched with 5% linseed oil; CAFL5%: obese rats fed the cafeteria diet enriched with 5% linseed oil. After verifying the normal distribution of the variables (Shapiro-Wilk test), the means of the six groups of rats were compared using a one-factor ANOVA test. This analysis was completed by the Tukey test in order to classify and compare the means in pairs. Means indicated by different letters (a, b, c) are significantly different ($p<0.05$).

Table A5. Total protein (g/l) and lipoprotein (mg/dl) levels in control and experimental rats.

Lots Parameters	Control rats			Obese rats			P (ANOVA)
	C	CL2.5% (%)	CL5%	CAF	CAFL2.5	CAF5%.	
proteins total (g/l)	34,36 ±7,52	38,38 ±5,45	34,25 ± 6,33	45,57 ±6,01	44,21 ±5,03	43,27 ±7,23	0,12 1
VLDL (mg/dl)	38±0,03	37±0,15	28±0,08	41±0,08	37±0,09	35±0,08	0,2 1
LDL (mg/dl)	56±0,05 [c]	56±0,05 [c]	57±0,04 [c]	10±0,02 [a]	64±0,01 [b]	63±0,01 [b]	0,01
HDL (mg/dl)	65±0,04[b]	60±0,02[b]	62±0,02[b]	71±0,04a	64±0,02[b]	64±0,01[b]	0,01

Each value represents the mean ± SD, n=10.C: control rats fed the standard diet; CAF: obese rats fed the cafeteria diet; CL2.5%: control rats fed the standard diet enriched with 2.5% linseed oil; CAFL2.5%: obese rats fed the cafeteria diet enriched with 2.5% linseed oil; CL5%: control rats fed the standard diet enriched with 5% linseed oil; CAFL5%: obese rats fed the cafeteria diet enriched with 5% linseed oil. After verifying the normal distribution of the variables (Shapiro-Wilk test), the means of the six groups of rats were compared using a one-factor ANOVA test. This analysis was completed by the Tukey test in order to classify and compare the means in pairs. Means indicated by different letters (a, b, c) are significantly different ($p<0.05$).

Lots Parameters	Control rats			Obese rats			P (ANOVA)
	C	CL2.5% FOR	CL5%	CAF	CAFL2.5	CAFL5%	
Liver (g)	21.19±1.43[c]	20.13±1.2[c]	20.52±1.7[c]	24.31±1.27[a]	21.21±1.09[b]	22.67±1.12[b]	0,001
Muscle (g)	5,59±0,57 [c]	5,36±0,34 [c]	5,31±0,27 [c]	7,43±0,41 [a]	6,37±0,38 [b]	6,44±0,27 [b]	0,01
Adipose tissue (g)	7.63±0.57[d]	7.25±0.11[d]	5.39±0.88[e]	19.08±1.90[a]	14.42±1.39[b]	10.26±1.6[c]	0,0001

Each value represents the mean ± SE, n=10.C: control rats fed a standard diet; CAF: obese rats fed a cafeteria diet; CL2.5%: control rats fed a standard diet enriched with 2.5% linseed oil; CAFL2.5%: obese rats fed a cafeteria diet enriched with 2.5% linseed oil; CL5%: control rats fed a standard diet enriched with 5% linseed oil; CAFL5%: obese rats fed a cafeteria diet enriched with 5% linseed oil. After verifying the normal distribution of the variables (Shapiro-Wilk test), the means of the six groups of rats were compared using a one-factor ANOVA test. This analysis was completed by the Tukey test in order to classify and compare the means in pairs. Means indicated by different letters (a, b, c) are significantly different ($p<0.05$).

Table A7. Total lipid content (mg/g tissue) of organs in different batches of rats

Lots	Control rats			Obese rats			*P ANOVA*
Parameters	**C**	**CL2.5% (%)**	**CL5%**	**CAF**	**CAFL2.5**	**CAFL5%**	
Liver (mg/g)	103,81±6,2^{c}	99,01±7,10^{c}	76,73±6,50^{d}	193,58±12,86^{a}	145,45±10^{b}	139,41±9,77^{b}	0,01
Muscle (mg/g)	72,33±5,49^{b}	73,96±6,99^{b}	72,91±4,58^{b}	81,83±4,43^{a}	82,83±5,59^{a}	82,5±3,99^{a}	0,01
Adipose tissue (mg/g)	236,68±15^{b}	136,71±11,37^{d}	126,31±13^{d}	339±21,21 ^{a}	218,25±13^{b}	182±14,37 ^{c}	0,01
Intestine (mg/g)	84,91±11,14	84,5±8,74	81,41±11,83	87,56 ±11,1	85,66±8,74	83,8±7,9	0,3 15

Each value represents the mean ± SD, n=10.C: control rats fed the standard diet; CAF: obese rats fed the cafeteria diet; CL2.5%: control rats fed the standard diet enriched with 2.5% linseed oil; CAFL2.5%: obese rats fed the cafeteria diet enriched with 2.5% linseed oil; CL5%: control rats fed the standard diet enriched with 5% linseed oil; CAFL5%: obese rats fed the cafeteria diet enriched with 5% linseed oil. After verifying the normal distribution of the variables (Shapiro-Wilk test), the means of the six groups of rats were compared using a one-factor ANOVA test. This analysis was completed by the Tukey test in order to classify and compare the means in pairs. Means indicated by different letters (a, b, c) are significantly different (p<0.05).

Table A8. Total cholesterol levels in the organs of different batches of rats

Lots	Control rats			Obese rats			*P* ANO
Parameters	**C**	**Cl2.5%**	**CL5%**	**CAF**	**CAFL2.5**	**CAFL5%**	VA
Liver (mg/g)	40,11±2,38^{b}	21,20±1,76^{d}	20,05±1,37^{d}	62,22±0,48^{a}	35,05±2,59^{c}	32,11±1,69^{c}	0,001
Adipose tissue(mg/g)	90,22±1,95^{b}	49,19±1,77 ^{d}	43,05±2,21^{d}	165,33±2,22^{a}	79,67±2,23 ^{c}	75,3±1,12 ^{c}	0,001
Muscle (mg/g)	23,14±2,04^{b}	19,87±3,47^{c}	18,54±1,91^{c}	38,67±3,45^{a}	19,7±1,32^{c}	19,03±2,89^{c}	0,01
Intestine (mg/g)	23,12±4,44	24,31±3,97	23,92±5,15	26,21±3,73	24,92±4,30	25,33±3,14	0,02

Each value represents the mean ± SD, n=10.C: control rats fed the standard diet; CAF: obese rats fed the cafeteria diet; CL2.5%: control rats fed the standard diet enriched with 2.5% linseed oil; CAFL2.5%: obese rats fed the cafeteria diet enriched with 2.5% linseed oil; CL5%: control rats fed the standard diet enriched with 5% linseed oil; CAFL5%: obese rats fed the cafeteria diet enriched with 5% linseed oil. After verifying the normal distribution of the variables (Shapiro-Wilk test), the means of the six groups of rats were compared using a one-factor ANOVA test. This analysis was completed by the Tukey test in order to classify and compare the means in pairs. Means indicated by different letters (a, b, c) are significantly different (p<0.05).

Table A9. Triglyceride content of organs from different batches of rats

Lots	Control rats			Obese rats			*P*
Parameters	**Controls (C)**	**Cl2.5%**	**CL5%**	**Obese witnesses**	**CAFL2.5**	**CAFL5%**	ANO VA
Liver (mg/g)	39,44±1,38^{b}	21,22±1,76^{d}	20,14±1,37^{d}	61,32±1,48^{a}	35,12±1,59^{c}	32,23±1,69^{c}	0,001
Adipose tissue (mg/g)	9,32±2,95^{b}	50,12±3,77 ^{d}	45,33±3,21^{d}	164,22±5,22^{a}	81,22±5,23 ^{c}	79,23±7,12 ^{c}	0,001
Muscle (mg/g)	23,12±1,04^{b}	19,9±2,47^{c}	18,7±1,91^{c}	35,22±3,45^{a}	18,9±4,32^{c}	18,44±5,89^{c}	0,01
Intestine (mg/g)	25,44±1,44	25,9±2,97	28 ,6±3,15	26,65±4,73	26 ,21±4,30	26,3±3,14	0,2

Each value represents the mean ± SE, n=10.C: control rats fed the standard diet; CAF: obese rats fed the cafeteria diet; CL2.5%: control rats fed the standard diet enriched with 2.5% linseed oil; CAFL2.5%: obese rats fed the cafeteria diet enriched with 2.5% linseed oil; CL5%: control rats fed the standard diet enriched with 5% linseed oil; CAFL5%: obese rats fed the cafeteria diet enriched with 5% linseed oil. After verifying the normal distribution of the variables (Shapiro-Wilk test), the means of the six groups of rats were compared using a one-factor ANOVA test. This analysis was completed by the Tukey test in order to classify and compare the means in pairs. Means indicated by different letters (a, b, c) are significantly different (p<0.05).

Table A10. Total protein content (mg/g tissue) of organs in different batches of rats

Lots	Control rats			Obese rats			*P* ANOVA
Parameters	**C**	**Cl2.5%**	**CL5%**	**CAF**	**CAFL2.5**	**CAFL5%**	
Liver (mg/g)	39,36±8,38[b]	21,83±4,76[d]	20,68±5,37[d]	60,92±4,48[a]	32,47±5,59[c]	30,13±7,69[c]	0,002
Adipose tissue (mg/g)	84,17±2,9[b]	71,22±3,77 [d]	70,27±3,21[d]	121,31±5,22[a]	70,37±5,23 [c]	69,05±10,12[c]	0,002
Muscle (mg/g)	62,45 ±3,04	61,11±3,47	59,83±1,91	63,37±3,45	63,85±4,32	50,89±5,89	0,19
Intestine (mg/g)	25,32±4,44	23,79±4,97	25,74±1,15	24,20±2,73	23,62±4,30	23,87±3,14	0,2

Each value represents the mean ± SD, n=10.C: control rats fed the standard diet; CAF: obese rats fed the cafeteria diet; CL2.5%: control rats fed the standard diet enriched with 2.5% linseed oil; CAFL2.5%: obese rats fed the cafeteria diet enriched with 2.5% linseed oil; CL5%: control rats fed the standard diet enriched with 5% linseed oil; CAFL5%: obese rats fed the cafeteria diet enriched with 5% linseed oil. After verifying the normal distribution of the variables (Shapiro-Wilk test), the means of the six groups of rats were compared using a one-factor ANOVA test. This analysis was completed by the Tukey test in order to classify and compare the means in pairs. Means indicated by different letters (a, b, c) are significantly different (p<0.05).

Lots	Control rats			Obese rats			*P* ANOVA
Parameter	**C**	**Cl2.5%**	**CL5%**	**CAF**	**CAFL2.5**	**CAFL5%**	
Liver µmol/g/min	41,67 ± 4,48[e]	69,58±5,59[c]	63,10 ± 5,37[c]	53,83 ±8,38[d]	87,39 ±4,76[a]	79,00±5,37[b]	0,001
Adipose tissue µmol/g/min	42,32±2,95[d]	45,21±3,77 [d]	51,23±3,21[c]	60,22±5,22[b]	90,01±5,23 [a]	88,5±3,12[a]	0,001
Muscle µmol/g/min	29,11±3,04[a]	31,03±3,47[a]	30 ,12±1,91[a]	19,4±3,45[c]	25,12±1,22[b]	29,33±3,89[b]	0,01

Each value represents the mean ± SE, n=10.C: control rats fed the standard diet; CAF: obese rats fed the cafeteria diet; CL2.5%: control rats fed the standard diet enriched with 2.5% linseed oil; CAFL2.5%: obese rats fed the cafeteria diet enriched with 2.5% linseed oil; CL5%: control rats fed the standard diet enriched with 5% linseed oil; CAFL5%: obese rats fed the cafeteria diet enriched with 5% linseed oil. After verifying the normal distribution of the variables (Shapiro-Wilk test), the means of the six groups of rats were compared using a one-factor ANOVA test. This analysis was completed by the Tukey test in order to classify and compare the means in pairs. Means indicated by different letters (a, b, c) are significantly different (p<0.05).

Lots	Control rats			Obese rats			*P* ANOV
Parameters	**C**	**Cl2.5%**	**CL5%**	**CAF**	**CAFL2.5**	**CAFL5%**	**A**
LCAT (nmol/ml/h	36,58±1,49[b]	31,41±1,4[c]	30,5±1,78[c]	42,25±3,15[a]	38,83±1,36[b]	35,25±2,03 [b]	0,002

Each value represents the mean ± SE, n=10.C: control rats fed the standard diet; CAF: obese rats fed the cafeteria diet; CL2.5%: control rats fed the standard diet enriched with 2.5% linseed oil; CAFL2.5%: obese rats fed the cafeteria diet enriched with 2.5% linseed oil; CL5%: control rats fed the standard diet enriched with 5% linseed oil; CAFL5%: obese rats fed the cafeteria diet enriched with 5% linseed oil. After verifying the normal distribution of the variables (Shapiro-Wilk test), the means of the six groups of rats were compared using a one-factor ANOVA test. This analysis was completed by the Tukey test in order to classify and compare the means in pairs. Means indicated by different letters (a, b, c) are significantly different (p<0.05).

Table A13. Activity of the hormone-sensitive lipase enzyme (HSL) in adipose tissue from different batches of rats

Lots	Control rats			Obese rats			*P* (ANOVA)
Parameters	**C**	**CL2.5% FOR**	**CL5%**	**CAF**	**CAFL2.5**	**CAFL5%**	

LHS (gmol/g/min)	280,12±18,38^{b}	170,33±14,76^{d}	168,45±15,37^{d}	360,34±24,48^{a}	245,54±15,59^{c}	169,23±17,69^{d}	0,001

Each value represents the mean ± SD, n=10.C: control rats fed the standard diet; CAF: obese rats fed the cafeteria diet; CL2.5%: control rats fed the standard diet enriched with 2.5% linseed oil; CAFL2.5%: obese rats fed the cafeteria diet enriched with 2.5% linseed oil; CL5%: control rats fed the standard diet enriched with 5% linseed oil; CAFL5%: obese rats fed the cafeteria diet enriched with 5% linseed oil. After verifying the normal distribution of the variables (Shapiro-Wilk test), the means of the six groups of rats were compared using a one-factor ANOVA test. This analysis was completed by the Tukey test in order to classify and compare the means in pairs. Means indicated by different letters (a, b, c) are significantly different (p<0.05).

Table A14. Serum vitamin C and erythrocyte glutathione levels and activity of the erythrocyte enzyme catalase in different batches of rats.

Lots	Control rats			Obese rats			P
Parameters	C	CL2.5% (%)	CL5%	CAF	CAFL2.5	CAFL5%	(ANOVA)
Vitamin C	12,19 ±0,24 ^{d}	21,41±1,02 a	20,07±0,52 ^{b}	6,18±0,40^{e}	14,66±2,04 ^{c}	15,65±0,50 ^{c}	0,01
Glutathione	3,25±0,14 ^{b}	3,69±0,05 ^{b}	4,90±0,17 a	1,19±0,08 ^{d}	2,58±0,21 ^{c}	2,89±0,34 ^{c}	0,001
Catalase	234,97±6,99 a	161,27±4,9 ^{c}	201±4,27 ^{b}	95,17±3,64 ^{f}	102,76±3,16^{e}	137,86±3,46^{d}	0,004

Each value represents the mean ± SE, n=10.C: control rats fed the standard diet; CAF: obese rats fed the cafeteria diet; CL2.5%: control rats fed the standard diet enriched with 2.5% linseed oil; CAFL2.5%: obese rats fed the cafeteria diet enriched with 2.5% linseed oil; CL5%: control rats fed the standard diet enriched with 5% linseed oil; CAFL5%: obese rats fed the cafeteria diet enriched with 5% linseed oil. After verifying the normal distribution of the variables (Shapiro-Wilk test), the means of the six groups of rats were compared using a one-factor ANOVA test. This analysis was completed by the Tukey test in order to classify and compare the means in pairs. Means indicated by different letters (a, b, c) are significantly different (p<0.05).

Table A15. Markers of plasma oxidative status in different batches of rats

	Control rats			Obese rats			
Lots Parameters	C	CL2.5% FOR	CL5%	CAF	CAFL2.5	CAFL5%	*P* ANOVA
MDA (pmol/l)	2,86±0,19 ^{d}	2,59±0,16 ^{e}	2,02±0,01^{f}	5,44±0,38 a	3,74±0,36 ^{b}	3,07±0,29 ^{c}	0,0001
Hydroperoxides (pmol/l)	8,9±0,39 b	7,19±0,15 ^{c}	5,42±0,29 ^{e}	11,8±0,25 a	6,4±0,22 ^{c}	5,86±0,19 ^{d}	0,001
Carbonylated proteins (pmol/l)	3,24±0,06 ^{b}	2,4±0,08 ^{c}	2,97±0,12^{b}	5,36±0,24 a	3,42±0,06 ^{b}	3,17±0,09 ^{b}	0,001
DIC (pmol/l)	2,96±1,37 ^{b}	2,42±1,01 ^{d}	2,38±1,10 ^{d}	3,81±1,43 a	2,50±2,36 ^{c}	2,43±1,91 ^{c}	0,001
Lipoprotein oxidation	1,62±1,37 ^{c}	1,55±1,01 ^{c}	1,40±1,01 ^{d}	2,83±1,01 a	1,93±1,01 ^{b}	2,09±1,01 ^{b}	0,001

Each value represents the mean ± SD, n=10.C: control rats fed the standard diet; CAF: obese rats fed the cafeteria diet; CL2.5%: control rats fed the standard diet enriched with 2.5% linseed oil; CAFL2.5%: obese rats fed the cafeteria diet enriched with 2.5% linseed oil; CL5%: control rats fed the standard diet enriched with 5% linseed oil; CAFL5%: obese rats fed the cafeteria diet enriched with 5% linseed oil. After verifying the normal distribution of the variables (Shapiro-Wilk test), the means of the six groups of rats were compared using a one-factor ANOVA test. This analysis was completed by the Tukey test in order to classify and compare the means in pairs. Means indicated by different letters (a, b, c) are significantly different (p<0.05).

Table A16. Liver oxidant/antioxidant status markers in different batches of rats

Lots	Control rats			Obese rats			P
Parameters	C	CL2.5% (%)	CL5%	CAF	CAFL2.5	CAFL5%	(ANOVA)

MDA (nmol/g)	4,37±0,51 ^{b}	1,66±0,3^{c}	1,42±0,3 ^{d}	5,90±0,81 ^{a}	2,63±0,40^{c}	2,28±0,21^{c}	0,001
Carbonylated proteins (nmol/g)	4,02±0,09 ^{c}	3,10±0,19 ^{d}	1,47±0,13 ^{e}	5,26±0,26 ^{a}	3,4±0,08 ^{b}	2,87±0,04 ^{c}	0,001
Glutathione (pmol/g)	0,51±0,05 ^{a}	0,42±0,04 ^{c}	0,41±0,05 ^{c}	0,89±0,08 ^{b}	0,63±0,03 ^{c}	0,61±0,06 ^{c}	0,01

Each value represents the mean ± SE, n=10.C: control rats fed the standard diet; CAF: obese rats fed the cafeteria diet; CL2.5%: control rats fed the standard diet enriched with 2.5% linseed oil; CAFL2.5%: obese rats fed the cafeteria diet enriched with 2.5% linseed oil; CL5%: control rats fed the standard diet enriched with 5% linseed oil; CAFL5%: obese rats fed the cafeteria diet enriched with 5% linseed oil. After verifying the normal distribution of the variables (Shapiro-Wilk test), the means of the six groups of rats were compared using a one-factor ANOVA test. This analysis was completed by the Tukey test in order to classify and compare the means in pairs. Means indicated by different letters (a, b, c) are significantly different ($p<0.05$).

Table A17. Markers of muscle oxidant/antioxidant status in different batches of rats

Lots	**Control rats**			**Obese rats**			***P* (ANOVA)**
Parameters	**C**	**CL2.5% (%)**	**CL5%**	**CAF**	**CAFL2.5**	**CAFL5%**	
MDA (nmol/g)	0,57±0,02 ^{d}	1,52±0,03 ^{b}	1,74±0,03 ^{a}	0,42±0,03 ^{e}	1,38±0,02 ^{c}	1,77±0,04^{a}	0,001
Carbonylated proteins (nmol/g)	0,33±0,04 ^{c}	1,24±0,03 ^{a}	1,29±0,04 ^{a}	0,33±0,04 ^{c}	1,05±0,08 ^{b}	0,98±0,05 ^{b}	0,004
Glutathione (gmol/g)	0,67±0,04	0,68±0,02	0,67±0,05	0,72±0,06	0,68±0,04	0,67±0,04	0,132

Each value represents the mean ± SD, n=10.C: control rats fed the standard diet; CAF: obese rats fed the cafeteria diet; CL2.5%: control rats fed the standard diet enriched with 2.5% linseed oil; CAFL2.5%: obese rats fed the cafeteria diet enriched with 2.5% linseed oil; CL5%: control rats fed the standard diet enriched with 5% linseed oil; CAFL5%: obese rats fed the cafeteria diet enriched with 5% linseed oil. After verifying the normal distribution of the variables (Shapiro-Wilk test), the means of the six groups of rats were compared using a one-factor ANOVA test. This analysis was completed by the Tukey test in order to classify and compare the means in pairs. Means indicated by different letters (a, b, c) are significantly different ($p<0.05$).

Table A18. Markers of oxidative/antioxidant status of adipose tissue in different batches of rats

Lots	**Control rats**			**Obese rats**			***P* (ANOVA)**
Parameters	**C**	**CL2.5% FOR**	**CL5%**	**CAF**	**CAFL2.5**	**CAFL5%**	
MDA (nmol/g)	3,04±0,24 ^{c}	2,53±0,6^{d}	2,33±0,5^{d}	4,67±0,26 ^{a}	3,88±0,5 ^{b}	3,38±0,08^{c}	0,001
Carbonylated proteins (nmol/g)	2,82±0,08^{b}	1,94±0,10^{d}	1,65±0,13^{e}	3,59± 0,05^{a}	2,22±0,11^{c}	2,11± 0,09^{c}	0,001
Glutathione (pmol/g)	0,44±0,02	0,45±0,03	0,42±0,03	0,48±0,03	0,48±0,03	0,47±0,04	0,1

Each value represents the mean ± SD, n=10.C: control rats fed the standard diet; CAF: obese rats fed the cafeteria diet; CL2.5%: control rats fed the standard diet enriched with 2.5% linseed oil; CAFL2.5%: obese rats fed the cafeteria diet enriched with 2.5% linseed oil; CL5%: control rats fed the standard diet enriched with 5% linseed oil; CAFL5%: obese rats fed the cafeteria diet enriched with 5% linseed oil. After verifying the normal distribution of the variables (Shapiro-Wilk test), the means of the six groups of rats were compared using a one-factor ANOVA test. This analysis was completed by the Tukey test in order to classify and compare the means in pairs. Means indicated by different letters (a, b, c) are significantly different (p<0.05).

Table A19. Markers of gut oxidant/antioxidant status in different batches of rats

Lots Parameters	Control rats C	CL2.5% FOR	CL5%	Obese rats CAF	CAFL2.5	CAFL5%	*P* (ANOVA)
MDA (nmol/g)	1,23±0,05	3,12±0,11	3,18±0,06	1,52±0,05	3,28±0,05	3,15±0,06	0,001
Carbonylated proteins (nmol/g)	0,43±0,03	1,94±0,03	2,23±0,11	0,54±0,02	2,31±0,06	2,26±0,04	0,001
Glutathione (Hmol/g)	0,59±0,06	0,57±0,04	0,50±0,06	0,52±0,05	0,53±0,05	0,58±0,05	0,1

Each value represents the mean ± SE, n=10.C: control rats fed the standard diet; CAF: obese rats fed the cafeteria diet; CL2.5%: control rats fed the standard diet enriched with 2.5% linseed oil; CAFL2.5%: obese rats fed the cafeteria diet enriched with 2.5% linseed oil; CL5%: control rats fed the standard diet enriched with 5% linseed oil; CAFL5%: obese rats fed the cafeteria diet enriched with 5% linseed oil. After verifying the normal distribution of the variables (Shapiro-Wilk test), the means of the six groups of rats were compared using a one-factor ANOVA test. This analysis was completed by the Tukey test in order to classify and compare the means in pairs. Means indicated by different letters (a, b, c) are significantly different (p<0.05).

Printed by Books on Demand GmbH, Norderstedt / Germany